# TIPS FROM THE TRENCHES

# TIPS FROM THE TRENCHES

## AMERICA'S BEST TEACHERS DESCRIBE EFFECTIVE CLASSROOM METHODS

**CHARLES M. CHASE, Ed.D.**

**JACQUELINE E. CHASE, M.A.**

TECHNOMIC
PUBLISHING CO., INC.

LANCASTER · BASEL

## Tips from the Trenches
a **TECHNOMIC**® publication

*Published in the Western Hemisphere by*
Technomic Publishing Company, Inc.
851 New Holland Avenue, Box 3535
Lancaster, Pennsylvania 17604 U.S.A.

*Distributed in the Rest of the World by*
Technomic Publishing AG
Missionsstrasse 44
CH-4055 Basel, Switzerland

Printed in the United States of America
10  9  8  7  6  5  4  3

Main entry under title:
   Tips from the Trenches: America's Best Teachers Describe Effective Classroom Methods

A Technomic Publishing Company book
Bibliography: p.

Library of Congress Catalog Card No. 93-60578
ISBN No. 1-56676-054-2

*To our sons, Brandon and Cade, with whom we get to learn all over again, and in special remembrance of our good friend and educator, Maxine Brumfield.*

IN any discussion of the profession of teaching, or in any examination of how the culture of that profession is infused into its members, the focus tends toward philosophical tenets, theoretical constructs, or research-based best practices. The impact of that body of knowledge acquired by teachers during the practice of their chosen profession is slight. This *teacher lore* has become an oral tradition similar to the oral tradition existing in more primitive societies. It is a collection of information that is extremely functional for those teachers who must exist in a challenging, often hostile, environment (just as the environment of primitive individuals is challenging and hostile). While the lore of the practitioner has not been widely considered of prime importance, it is often the only lifeline beginning teachers have to hold on to as they attempt to survive their difficult initiations.

The authors have included the uncommon sense of seasoned, professional educators in an easily readable and user-friendly text. This book is a valuable resource for the beginning teacher, the veteran practitioner, or the university teacher-educator. Each time I turn the pages of this text I find practical suggestions for effective instructional practice. Suggestions for enhancing student motivation, working with parents, or managing child behavior, all are encompassed in the parameters of this work.

This text attempts to bring together the polar positions of those in the field of education, to serve as a bridge between preservice instruction and professional practice, to validate that *all* educators have the experience, knowledge, and the ability to contribute to the profession. To contribute to and improve upon our profession should be the prime goal of each of us. This text achieves both these goals, provides the basis for instructional improvement, and ultimately promises a positive impact on those most directly concerned with our profession—the students of today and tomorrow.

J. M. Blackbourn
*University of Mississippi*

THIS is a book for educators and education administrators. More importantly, this is a book based totally on the input of educators from throughout the United States. Further, these educators have been individually identified as being *exemplary* teachers by their building principals. These are the people in the trenches, the men and women with chalk dust on their clothes. These are the teachers who contribute so significantly to the growth and development of our children.

The exemplary teachers who have contributed to the content of this book are sharing what works for them regarding the top ten problems faced by entry-level or first-year teachers, as indicated by first-year teachers in numerous research studies.

Although it is probably the first-year teacher who is most in need of this valuable collection of classroom wisdom, this same material will be well-received by early-career and veteran teachers. Additionally, school administrators such as superintendents, principals, curriculum directors, counselors, and others will find the content to be most insightful and useful in their day-to-day goings-on.

As you read through this book, you will meet ninety-five exemplary teachers from throughout the United States. Some teach in our biggest cities and largest school districts. Others have classrooms in our most remote rural areas. Some conduct classrooms at the elementary level, and some at the secondary level. Because readers of this book will come from a wide variety of classroom settings, and because we all pretty well agree that any information is most valuable when it is shared — especially with an opportunity for continued discussion — this book has been designed so any reader who chooses to contact a teacher who has contributed to the content of the book may do so in almost all cases.

When each of the contributing teachers provided their input on their chosen problems, they also indicated whether or not they were willing to make their names and addresses known to the reader. With very few exceptions, the contributing teachers have willingly provided such information, and by so doing, welcome your follow-up inquiries. The process for contacting any of these teachers is very simple. After each entry in each chapter, the identity of the contributor is provided. If the entry contained material that intrigued or excited you, and you would like to discuss the content or related items more in depth, refer to the Appendix at the end of the book. Go first to the appropriate section, elementary or secondary, and then find the teacher of your choice who will be listed in alphabetical order in that section. It is there you will find the current address of the teacher to whom you can send your questions.

The content of this book, plus the opportunity to network with the contributing exemplary teachers, is intended to provide a focus and a knowledge base of *best practice* or *what works* as opposed to theory or research. Without doubt, the most effective classroom and educational system will be founded on a blend of research and best practice. Research findings are generally quite accessible, but the same cannot usually be said for what the classroom practitioners have found to be best practice — what works. Thus the design, focus, and intent of this book.

Because of the constraints of time and extent of material, the contributors to this book were selected by use of the following research methodology and sampling design.

- Using the most current directory of schools for each state, the two largest cities for each state were identified.
- From each city, one public elementary school and one public secondary school were randomly selected.
- Within each state, four additional cities were randomly selected and from within each of those cities, one public elementary school and one public secondary school were selected.
- The principal of each of these schools was contacted to identify an exemplary teacher, who in turn provided the content for this book.

Accordingly, the contributing teachers and their input cover most conceivable settings and conditions.

The principals who selected an exemplary teacher from their faculty were simply asked to do just that — select an exemplary teacher. In keeping with an outcomes-based concern (as opposed to a process), what really matters is how well one performs in action, not what a list of credentials or criteria may suggest one is capable of doing. In most cases, no one person in any given school should be better qualified to identify an exemplary teacher, a top-notch in-action teacher, than the school's CEO, the principal. Therefore, so as not to distort any principal's decision regarding who he or she perceived as being exemplary, no potentially restrictive or skewing criteria were imposed on the selection process.

A critical part of all teacher preparation programs, be they traditional residential programs at institutes of higher learning, or the more recent collaboratively based alternative certification programs, is the time devoted to site-based participation by the intern teacher. All too often, the connection between the classroom dimension of the intern's training and the reality of the school site is somewhat indistinct. It is the purpose of this book to help bridge that very gap by providing the following:

- successful classroom strategies as evidenced by *practicing teachers*
- input from *many* successful teachers, rather than the perceptions or beliefs of a single author
- time-tested, practical school and classroom management information
- a savvy book of knowledge for teachers and school administrators who want to improve themselves, their schools, and enhance academic achievement in the classroom
- a professional book for teachers and administrators
- a valuable addition for professional libraries

The editors wish to provide acknowledgement and a special thank-you to Margaret Talbot for the long hours and commitment to task and excellence in the preparation of this book.

ASK any teacher, observe any classroom, or read the research concerning the problems and frustrations common to almost all first-year or early-career teachers, and you will discover a list of items or issues that generate what teachers often refer to as *September dread,* or *baptism by fire.*

This book addresses these points. If ignored or taken lightly, these things will become primary obstacles to success in the classroom. This applies not only to entry-level or early-career teachers, but to veteran teachers as well. As revealed by a compilation of eighty-three research studies, the specific concerns are:

- classroom discipline
- motivating students
- dealing with individual differences in students
- assessing student work
- relations with parents
- classroom management (organization for teaching)
- insufficient materials and supplies
- dealing with problems of individual students
- heavy teaching load with insufficient preparation
- relations with colleagues

These concerns represent the perceived problems of beginning teachers. The structure of this list is based on a rather exhaustive study done in 1984 [Veenman, S., *Review of Educational Research,* 54(2)].

If our intent is to help teachers achieve success in their classrooms, we should not only be addressing the above list of issues, but we should also be doing so from the proper perspective—the practicing classroom teacher. Unfortunately, too many of our teacher education programs have been giving too little attention to the site-based, practitioner-oriented aspect of training. This, however, is changing as we continue to recognize this deficiency. For example, major accrediting agencies in the field of education have made significant shifts away from the pedantic classroom in a university toward the real classroom in the public schools. In addition, the current focus on site-based management contributes to this shift in focus. Finally, the most recent effort to move from process (curriculum) assessment to outcomes-based assessment, is yet another way in which the minimized focus on site-based, practitioner-oriented information is being revealed. Thus, the reason for *Tips from the Trenches*. Here is a vehicle for entry-level and early-career teachers by which they can be apprised of successful and implementable strategies which attend to the most common occurring problems in teaching, as suggested by those who know best—the exemplary practitioners.

The book has been designed in a user-friendly format to provide a quick and usable reference for teachers and school administrators, whether new on the job or veterans. First, the book is divided into two major sections—one for educators at the elementary level and one for educators at the secondary level. Each of these major sections is further divided into ten chapters representing the top concerns or problems, as evidenced by numerous teacher encounters. Within each major section, the chapters are ordered to represent, in descending order, how the contributing veteran exemplary teachers responded to them. In other words, the higher the number of responses, the higher on the list the particular problem was ranked.

The reader is encouraged to consume the content in a beginning-to-end fashion, thereby getting a comprehensive overview of the entire spectrum of the public school classroom. However, should a reader have a singular interest at either the elementary or secondary level, that interest can be easily and readily accessed by simply consulting the Table of Contents in the front of the book.

Each contributing teacher's input is his or her personalized response to several of the problems delineated earlier. The responses are in the form of "Here's how to deal with problem #____."

*Tips from the Trenches* offers the reader 452 field-tested strategies regarding "what works" for ninety-five exemplary teachers from forty-five states. We hope that you, whether teacher, administrator, or other education professional, will find many of these strategies new, exciting, and useful for your particular educational setting.

# ELEMENTARY EDUCATION

Classroom teachers, whether elementary or secondary, obviously share the ten problems that *Tips from the Trenches* focuses on. However, it is clear from the responses that elementary and secondary teachers rank their challenges differently. Therefore, the order of the chapters in Section One—Elementary Education, proceeds differently than it does in Section Two—Secondary Education. The challenges, in descending order of importance, according to the contributing elementary teachers, are:

- relations with parents
- classroom discipline
- motivating students
- classroom management
- insufficient materials and supplies
- dealing with individual differences
- heavy teaching load with insufficient preparation time
- relations with colleagues
- problems with individual students
- assessing student work

Along with each teacher's input, in most cases the name, grade level, and state of the teacher is also provided. Should the reader choose to communicate further with the teacher regarding the information given, the full mailing addresses of all named elementary teachers will be found in Appendix A.

## CHAPTER 1

# Relations with Parents

CONSIDER the fact that out of all the problems the veteran exemplary teachers responded to, they provided more responses to parent relations than any other. It would appear then, that it would be in the best interest of any entry-level or beginning teacher to take this particular issue very seriously.

Relations with parents, or for that matter, even giving thought to parents, hasn't always been a concern of teachers. But things have changed considerably. The administrative strategies that are espoused in our schools today include things like site-based management and a variety of collaborative designs. In most cases, parents are not only considered to be an integral part of these administrative structures, they are strongly encouraged and openly invited to be such.

It has taken us quite some time to discover what the veteran exemplary teachers will tell you in this chapter—parents are a necessary member of the team that ultimately does its best to ensure the best school experience possible for a parent's child. Further, the personal strategies shared in this chapter reveal a few widely endorsed beliefs as well as some unique approaches to enable you to deal with relations with parents head-on. You'll discover that good communication is worth its weight in gold. You'll also discover that you will probably encounter positive results when you involve parents in your classroom, get to know them as soon as possible, treat them as no less than a working partner, and stay in touch with them throughout the year.

The following suggestions provided by veteran exemplary teachers will expand upon these basic ideas as well as suggest several specific techniques. Have you ever heard of the "Oreo cookie" technique? Try it, it might be just what you need.

Good communication is the key. Being a first-grade parent twice before ever being a first-grade teacher, I realize that good communication is the key to a good working relationship between teacher and parent, one that the child will get maximum benefit from. One way of doing this is through a monthly newsletter. In this newsletter, include birthdays, units being studied, future units, supplies that will be needed, requests for volunteer help, "thank-you's" to parents who have helped in some way during the month, "quotable quotes" the students have made, skills being covered, and ways parents can help at home with these skills.

When parents do things for us, we send a personalized thank-you poem. I use the same poem each year, changing the names to include the new class. Most importantly, always end a teacher-parent conference by saying something positive about the student. Never end on a negative note.

Charlotte L. Hargett, *Grade 1, Mississippi*

---

On registration day, my parents and I discuss our school's kindergarten curriculum. I give each parent a copy of the core objectives that his/her child must master.

Communication with parents is very necessary. Suggested techniques include: progress reports in addition to report cards, informal notes on papers, newsletters giving information such as special events, requests for help on certain units, holiday dates, and, requests for parent conferences when necessary.

I encourage my parents to be involved with their children's school work. Sometimes parents visit my classroom to observe.

M. Faye McDonald, *Kindergarten, Mississippi*

---

Send a schoolroom newsletter home at least twice a month. This should be typed neatly and without errors and can be put on a patterned-lined paper. It should relate what units we are working on in all areas and what concepts the children should have developed. Use the newsletter to thank any parent who sent afternoon treats, to remind parents of upcoming events, and to relate any humorous happenings in the room.

Have parents come into the classroom to read orally or present a lesson, such as making nonbake cookies.

Have a preschool screening for each child. The parents must bring the child and watch as you screen him. This shows the parents where areas of concern may develop and you meet the parents on a one-to-one basis. This is the time to have the parent fill in a sheet of information about the child's background.

Faye Karna, *Kindergarten, North Dakota*

---

When having a parent conference, try to make a practice of telling parents something positive about their child before discussing a negative issue. This often helps parents to accept problems and to be cooperative in solving them.

When weekly communication with parents is necessary, the following method has proved successful. A weekly work folder is sent home. Upon receiving the child's work, parents sign a dated sheet which also has a place for parent/teacher comments. This tactic keeps communication open between parents and teachers.

*Anonymous, Elementary, Rhode Island*

---

Begin getting to know your parents the first week of school with a phone call to say, "Hi, I'm your child's teacher and I'm happy he's in my room." Telling parents one good thing their child has done seems to let them know you are both on the same side. I end my call with a "be sure to let me know if there are any problems" message and that opens the lines of communication.

*Marilyn K. Emmons, Grade 1, Montana*

---

In an attempt to have my students become more involved with our parent-teacher conferences, my students make a parent folder. Included in this folder are: (1) a page where the students write greetings to their parents or whoever attends their conferences in their behalf, (2) a page where they list or write about their favorite attractions in our room, and (3) most importantly, a page where the parents write their children special positive notes about each child's work. We then read these "love notes" the following day. Talk about a self-esteem booster! Oftentimes those are the only positive strokes these children receive from their parents.

*Barbara Herman, Grade 1, Oklahoma*

---

Encourage parent participation in your classroom. They can do many things to make your job easier, and most of them become more sympathetic to the problems of education if they are in school regularly so they know what is going on. I have had parents do such things as listen to kids read, check papers, run ditto copies, test children on sight word lists, help with creative writing projects, prepare programs, and other things.

Keep in touch with parents about their children's progress and behavior through phone calls, notes home, etc.

*Anonymous, Grade 4, Utah*

---

Parents are partners in the learning process of children. There is clear empirical evidence that family members can actually determine both what and how much children from their family unit learn; therefore, keep parents informed about topics relevant to your particular class.

Let them know how they can be active participants in the education of their children. This might include anything from tips or suggestions on helping children at home to classroom visits.

Julie Abell-Victory, *Grades K – 12, Pennsylvania*

---

I establish the climate of good rapport at the beginning of the year. At back-to-school night I not only talk to the parents, but hand them a packet of helpful information on how they can work with me and their child. I encourage drop-ins, phone calls, etc. Throughout the year I send home Friday folders with a couple of comments from me, and I send ''Good News'' telegrams relaying something positive about each child. These telegrams go out every week to a few kids at a time. I use phone calls and conferences all year to help with discipline problems. I contact the parent immediately to help the child before the problems get worse. I hold two luncheons or informal '' snack'' get-togethers throughout the year for chatting with parents.

Karen Morgan, *Grade 6, Idaho*

---

Let parents know something specific you like about their child. For example, '' She has a great sense of humor.'' '' I enjoy talking to him at recess about his fun weekends.''

Let parents know right away – up front – that you *like* their child.

Let the parents know things you'd like to work on with their child and enlist their support.

Ask parents for ways you can help them with problems or concerns they may have with the child.

Barbara Baker, *Grade K – 6, Hawaii*

---

First of all, don't think of this as a problem. It doesn't have to be. As a beginning teacher in a new community, I sent letters to the parents of my students about a month before school started in the fall. I introduced myself to them, told them about my background, told them what I hoped to accomplish that school year, etc. Most importantly, I invited them to feel welcome at school anytime. If parents feel that you value their opinions and suggestions, they'll support and help you. Get them involved in their child's education.

Nancy Mahloch, *Grade 6, Nebraska*

---

Parents want the best for their children so keep them informed by (1) monthly newsletters to inform them of topics you'll study for the coming month, (2) "Good News" notes through the mail (occasionally), (3) asking for parent input or suggestions in helping their child meet success, (4) thanking parents for their support and involvement, and (5) including parents in as many activities as possible.

Barb Parmenter, *Grade 4, Michigan*

When you deal with parents always be completely honest about what you see happening with their child. Be sure to listen to what they have to say about their child; after all, they've known that child longer. If you have something negative to discuss, try to use the "Oreo cookie" technique—tell them something positive at the *beginning* of the conversation and at the *end*, sliding the negative in the *middle*.

Sharon Kelly, *Grade 3, New Jersey*

Keep parents well-informed. Make contact early in the year, just to introduce yourself. At open house, make parents aware of your expectations for them as parents. Then just keep parents aware of positive as well as negative progress through phone calls, notes, or home visits.

Lesa Carroll, *Grade 5, Missouri*

Remember that the parents might be intimidated by the school. Don't ever question their parenting skills aloud. Be supportive of them, and present yourself as a partner in their child's education. If you need to tell a parent something about their child that they might not want to hear, sandwich it between two positive qualities you have observed in their child.

Christine Shepard, *Grade 2, Utah*

A unique way to build positive relations with parents is to use a newspaper. Most newspapers sponsor contests throughout the year. Let the students enter *any* and *all* of the contests, and they will win. Last Thanksgiving the students in grades four, five, and six entered a contest entitled "Spend the Holidays at the Midland Hilton." Twenty of the students were chosen as winners, and twenty families from this very small town spent the Thanksgiving holidays at the Hilton. Students learned, parents participated in the outcome, and the school was the hero. *Go for it!*

Shirley Spear, *Grades 4 – 6, Language Arts, Texas*

Meeting parents personally, as early as possible in the year, gives you a chance to establish a good relationship with them. I have an open house in my classroom so parents can see the surroundings and materials we use as well as talk with me informally. I stress the partnership we have in the education of their child and ask them to contact me with any questions or problems which might arise.

Shirley H. Goodemote, *Grade 1, New York*

---

Each day during the time your children are washing hands and preparing for lunch, have them copy the day's homework assignments in an 8″ × 11″ spiral notebook. Request a parent signature daily. The signature means the parent knows what is being taught and has checked the assigned work. Communicate personal notes in this same notebook. They can be congratulatory ones or ones of concern. For every phone call you make relating to a problem, make a matching one about an accomplishment. Request a parent signature on all major tests.

Betty Robeson, *Grade 3, Maryland*

---

Honesty is the very best technique in sharing a student's problem with parents. The teacher must worry about twenty-five other students; however, the parents have only one interest—their child. Keep this in mind! Offer both positive and negative comments/problems about the child. Look for something good to say. Invite parents to visit your classroom and share in the resolution of the problem. Establish an open-door policy.

Helen M. Martin, *Grade 4, Georgia*

---

It has been proven that the more interest the parents show, the harder their children will work. In that case, we as teachers need to have a good bonding with parents. Let them know that you are there for them and that *together* we can provide the best possible education for their children. Encourage parents to keep in contact with you by writing notes, telephone calls, or by stopping in. Remember—*together we can.*

Joe F. Coleman, *Grade 4, Ohio*

---

Without parent cooperation in dealing with behavioral and/or learning problems, you'll go nowhere fast. It's going to take more than keeping the communication lines open. Try to establish a positive relationship with all parents before problems develop—do it early in the year. As soon as you detect a problem that appears to be heading in a chronic direction, contact the parent(s). You must garner the support and backing of the parent(s) to move in a positive direction in solving, or at least improving, behavior and/or learning problems.

*Anonymous, Grade 4, Michigan*

---

I call every parent the first week. We discuss goals, rules, etc. in a positive way. Thereafter, I call about every other week. I always emphasize the positive, but I also discuss any problems at school, and I respond to problems on the home front. I have learned to be brief!

*Anonymous, Grade 6, Alabama*

---

Parents can be a teacher's most valuable resource. My advice to new teachers is to get to know the parents of your students as the school year begins. Early conferencing with parents, two weeks into the school year, can provide goal-planning time, information and parent insights, and a getting-to-know-you time.

Parent involvement *during the school year* is also extremely important. Ask parents to get involved in their child's education by providing parents the opportunity to participate. For example, a parent who is a nurse may speak to the children about health issues. A parent who works in the home may make learning center activities. Parents may volunteer to be special ''mystery readers'' to read to the children during story time. Parents who become actively involved will help to instill the importance of education in their children.

Becky Ann Erickson-Danielson, *Grade 1, Minnesota*

---

I feel that I establish a rapport with my students' parents, and I believe that I become talented in communicating with them when I learn to drop my guard so that I am able to view the parents as facilitators, not intruders (enemies). Most parents' questions are motivated by a need to understand their child's position in his/her learning environment. Even when a question appears to be a lack of trust in my ability/actions, such doubts are easily turned around when I make it evident that I am indeed listening to the parents' view (translated: their child's needs). Here is the key: first, listen and respond to what the parent wants you to hear (even when you disagree), repeat (confirm) what you have heard the concern to be, then offer information which you deem relevant in effecting a positive behavior or academic improvement.

Also, keep the parents well-informed. I send a weekly report for each student. I send daily reports for students who do not respond to my classroom management plan, and I respond immediately to all notes, phone calls, and requests for conferences.

Pauline McCollum, *Grade 2, Florida*

---

Make frequent contact. Notes are good, but phone calls are better. Try to have a parent-teacher conference with all parents as soon as possible. Be certain to contact parents when children are especially good, not just when there are behavioral or academic problems.

Kathryn Maloy, *Grade 2, Florida*

---

Communication is the key word here. Let parents know what their children are learning in the classroom through weekly or monthly communication. Teachers and/or students can write the newsletter. When a student has done something great or neat, write a positive note home. Invite parents to speak to the class about their jobs, hobbies, and interests. Encourage them to volunteer regularly in your classroom. Children love to read aloud to an adult and to share their story writing with them. Invite them to chaperone field trips and help with book fairs. Extend invitations to plays, teas, and other special events. Let them know you welcome them and value any assistance they can give to their children's school.

Sharon Conley, *Grade 2, Maine*

---

Establish a working relationship with parents right away. Have parents in to a group parent meeting early in the school year. Explain your program, your philosophy, and your expectations. Let them know your goals for the class, and your goals for the education of their students. Cover such areas as school policies, learning programs, and how they may help their child, both in and out of the room.

After the first meeting, follow-up *often* with *each* parent. Keep them informed of student achievements or areas of concern. Encourage their input and feedback—they know the child best. Make parental contact a routine of your teaching, whether by note or phone.

Finally, remember positive contacts lay a foundation for good communication, even if a serious problem may develop. It will be much easier to approach a problem if the parent knows you from previous contacts.

Sharron Baird, *Grades 2 – 4, Oregon*

---

It is important to remain open and honest at all times. I invite parents to come into my room any time during the day. When we do any class plays or other special activities, I invite parents to come. When I find that certain parents have a special talent, or have traveled to any area that we are studying, I invite them to be my resource persons. I send positive notes home or write notes or comments on papers I correct so parents can see I'm taking a special interest in their most prized possession. It's good to make phone calls to parents, but don't limit them to negative times for discipline reasons. During parent-teacher conferences, ask the parents for their support and emphasize that their child's achievement will be enhanced if there is a team approach of child-parent-teacher. Remove any one of the three parts and there will be an incomplete team.

Frankie C. Bly, *Grade 5, Minnesota*

---

This can be a complex issue. In the beginning of the year, I send a letter "Welcome to Room. . . ." Then, in October, I send a letter home asking the parents if they would like a brief statement about their child's progress. Parents seem to respond well to this. I schedule conferences at their convenience, within reason. I express concern over their children, always beginning with the positive first, then the negative. Toward the end of the conference I review everything and try to end on the positive. I also let the parents know that if I can help them with anything to please let me know. You need to let them feel at home. It is important that meetings with parents be nonthreatening.

Debra L. Mann, *Grades K – 5, Special Education, Massachusetts*

---

At the beginning of the year, we make an attempt to contact all our parents, introducing ourselves, sharing our philosophy, goals, etc. We encourage parent involvement through chaperone experiences and helping with classroom and after-school activities. Parents and students are given the opportunity to participate in YMCA night, school dances, etc.

Pam Smith/Terry Carter, *Grade 6, Illinois*

---

Be sure to keep in contact with parents. Conferences face-to-face are the best. Phone calls are helpful on a weekly basis with certain students. Always have an agenda you are going to discuss at a conference. Be sure to document your conferences. Parents like to get positive information about their children, so don't only make contact when there are negative issues. Parents want to be involved. Call them for help with theme units, holiday activities, field trips, and any special activity. The key is to stay in contact with parents.

Dorothy Hartson, *Grade 6, New Hampshire*

In the beginning of the year I have a "parent social." Invite the parents to an informal discussion of the fifth grade, complete with refreshments. Keep parents informed of their child's performance, both good and bad, throughout the year. Allow them to share their hobbies, collections, and experiences. Be honest with parents. Let them know you are available for questions or concerns. Most important, let them know how excited you are about teaching their sons and daughters.

Eileen Bruton, *Grade 5, Wisconsin*

I believe parents must be partners in the education process. For the past seven years, I have begun my school year with a parent open house. I explain my teaching philosophy, style, and goals for the year. I also give parents information on age-appropriate behaviors, what kinds of responses—emotionally, socially, and academically—they can expect to see.

I have an open-door policy in my classroom; parents can come in at any time. No appointments are necessary. I utilize parents to help with individual student assessments, enrichment and remediation activities, art projects, field trips, and help preparing daily materials.

I do another open house the third quarter of the school year to prepare parents for what they can expect the following school year.

Letters go home every two to three weeks explaining our class activities, needs, etc. Parents are my greatest resource.

Mary Ann Gildroy, *Kindergarten, Montana*

You should encourage a parent to share how she/he handles problems and successes for the child at home. Always be honest with the parent. Discuss the student's strengths and weaknesses and offer suggestions. *Always* try to end each parent conference on a positive tone.

Velma Bell, *Grades 3 – 5, Learning Disabled, Georgia*

Let parents know right from the start that they are an important part of their child's success in school. My closing on all letters to my parents is: "Your partner in education." I keep my parents informed of what is going on in the classroom, and how they can help their child by sending home newsletters. I try to have four parent meetings a year so the parents can meet their child's classmates' parents and find out about things that are happening in our classroom. I try to make my first contact with every parent positive, so they are willing to work with me if a problem arises.

Marcia Stirler, *Grades K, 3, Iowa*

When meeting with a parent I always try to start out the conference with two or three positive attributes about their child. Sometimes it's difficult, but with good preparation that can be achieved. Once I have praised their child, they are much more open to listening to me about problems, and are more willing to work with me on a solution. Even the parent whose child causes no problems needs to hear from time to time that things are O.K. I send home ''good notes'' about once a week and try to get to all parents at least once a semester.

Ron Baysinger, *Grade 6, Indiana*

Parent relations must always be a top priority for teachers. Teachers need the help of parents. *Be positive* in parent conferences. Tell parents the good behaviors and the good test scores first, then bring in the problem and ask for their help. Call parents in late afternoon and early evening, and make it a happy call to the home to praise a student, check on homework, etc. Make it brief and to the point.

Penny Parnell, *Grade 3, Alabama*

Be open and friendly with your parents. Invite them to come visit your classroom any time that is convenient for them. Make sure they know that you are available to them for questions, help, and advice if they should need some. Be positive about their child. Capitalize on the good things their child can do, and approach weaknesses with tenderness and caring. Never put parents down, or their parenting techniques. Suggestions for improvement, or help to be given to a child, should be given diplomatically and with concern and understanding. Try to be aware of the home situation so that you are not putting unnecessary pressure on an already tense situation. *Be understanding* of their situations and their children.

Daria Ann Wood, *Kindergarten, Wyoming*

Communication is the key to good relations with parents. Have a written note prepared to go home the first day outlining your expectations for students, your grading policy, the most convenient time to contact you at school, and a request for parental assistance. Follow up with newsletters (a great student project), computerized grade notices, and phone calls. Remember to write thank-you notes for field trip drivers, party treats, clerical assistance and so on.

Cay Spitzer, *Grade 6, Colorado*

I write a letter to the parents before school starts, asking them to describe their child in a letter, and to share anything that will help me to get to know their child. The parents are also invited to help out in the classroom, e.g., read to the children or listen to the children read. I want the parents to feel their input is valuable and welcome. Each quarter every parent is called to update their child's progress. Communication is a must.

Mary Geren-Saggau, *Grade 1, Nebraska*

## CHAPTER HIGHLIGHTS

Building a strong, positive relationship with your students and parents is a key building block. Remember, no matter how old their child is, most parents are concerned and want to know what's going on in the classroom.

It is very important to make parental contact as soon as possible in the new year. Newsletters, conferences, and phone calls keep parents in touch with the classroom. Use your parents as much as possible — in the classroom, for field trips, or help in preparing materials. A follow-up thank-you note will keep your parent worker coming cheerfully back to help.

Communication about the student's progress should be positive as well as telling the negative. Let parents feel free to drop in on the classroom.

## CHAPTER 2

# *Classroom Discipline*

CLASSROOM discipline is one area of concern that almost every teacher *and* school administrator will have to deal with on a daily basis. The issues range from small, pesky interactions with students, to the much more serious problems that periodically make headlines in major newspapers across the country.

Teachers and administrators take classroom discipline very seriously. In fact, some teachers will tell you that it is the key to a successful teaching career. Classroom discipline is also one of the facets of teaching that not only has everyone's attention, but has everyone in constant pursuit of what the latest findings are. Newly published texts, on a regular basis, attest to this as does the typically standing-room-only response to a session on classroom discipline at teacher conferences.

In the strategies and suggestions that follow, you will become sensitive to several common threads that seem to tie most of the responses together. In other words, there appear to be several postures and procedures that exemplary teachers have found to be dependable. Things like being *consistent* in your application of discipline, no matter what it is. Also, it is important to have your classroom rules *visible*. You need to be *fair*, you need to have your students *involved* in the development of the rules, and the rules should be *concise* and *established very early* in the year. But these are just a few of the many worthwhile, and more importantly, field-tested, classroom discipline methods that teachers have offered. Consider the following for possible use in your classroom.

---

An old saying says that first-year teachers should not crack a smile until after the first quarter. That's a little drastic. The key is to be fair, but firm and consistent. Assertive discipline has worked well for me. I've also had success with a point system. Involve the students in setting up the rules and consequences. If they have a feeling of ownership, they feel more responsibility for following those rules. Make them responsible for their behavior—they choose to follow or not follow the rules, therefore choosing the consequences. When setting up the consequences, don't just have negative ones. Include rewards and positive goals. It's also a good idea to share your discipline plan with parents, so they are informed how problems will be handled.

Nancy Mahloch, *Grade 6, Nebraska*

---

Classroom rules should be written in a positive way. Students need to have some input in the establishment of rules and consequences. Consistency is the key. Another important element is the establishment of a positive atmosphere. Students must feel they will be listened to. We attempt to talk about problems, use live examples, use humor, and have group discussions.

Pam Smith/Terry Carter, *Grade 6, Illinois*

---

In the beginning of the year I make this one statement, "For one year, we're going to be a family. We may get upset with each other, however, we need to help each other out." That seems to help break the ice.

When I discipline my class I use three baseballs; the concept being "three strikes you're out." If the students achieve three baseballs, they are not able to participate in outdoor events/indoor games which occur on Fridays. Instead, they can choose a book to read. This system is built into my program. For me, it works well. Also, I discipline with what I call "tough love." I never yell. However, I put more stress in my voice. I may give a look or may count to three or just wait until it is quiet. When I'm explaining something and stop for a particular reason, everyone knows why. I balance my discipline with positive reinforcement. I feel that the key is discipline that's appropriate, with positive reinforcement.

Debra L. Mann, *Grades K – 5, Special Education, Massachusetts*

---

I use Lee Cantor's assertive discipline model. I keep my rules to a minimum, and I have them posted in full view for my students. My three basic classroom rules are, (1) bring all materials to class and be ready to work, (2) listen when the teacher or a classmate is speaking, (3) listen to and follow all directions. As I'm teaching and I detect a problem area developing, I walk around the room and stand next to the person or persons causing the problem. My physical presence usually corrects any inappropriate behavior.

Frankie C. Bly, *Grade 5, Minnesota*

---

If done correctly, this never has to be a problem at all! The necessary foundation for classroom expectation must be clearly established right at the beginning of the year. Having the class help to determine the expectation allows them ownership for the success of the classroom program. As they establish the guidelines for behavior, they are implying accountability.

Keep the rules basic and achievable. If you can stand a loose environment, allow quiet talking. If you need a structured environment, build this into your class rules. Determine the consequences for breaking a rule.

Once established, by whatever means, the key to a successful classroom discipline program is consistency. Follow through each time expectations aren't up to par. Be fair and remind the student of what the expected behavior is. In addition, when behaviors are being met, reward the class as a whole. Praise and some small recognition of correct behavior are great preventive measures toward success of a classroom discipline program.

Sharron Baird, *Grades 2 – 4, Oregon*

---

Set a tone first. Tie a load down tight then let it wiggle loose. You can't be a friend then a disciplinarian, but you can loosen later. Give your students material that will challenge them. This stops a great deal of problems.

William Andrews, *Elementary, Massachusetts*

---

The key to classroom discipline is consistency. From day one, make sure the children know what to expect from you. We do a lot of discussion of the classroom rules and we even do role playing—as we are learning social/cooperative learning skills at the same time. I present them with a simple acronym that stands for our rules:

R.E.S.P.O.N.S.I.B.L.E.
R = respect others
E = exercise self control
S = share materials
P = prepare materials
O = on time!
N = nice behavior
S = smile!
I = invite success
B = be an active listener
L = leave room only with permission
E = encourage yourself and others
(Note that these are stated in the positive.)

Let's take "E" for example. . . . If someone *chooses* not to exercise self control, then they have also chosen to take a "time out." The children know their actions are what determine mine. This puts things more in their hands.

I also have to mention our credit card system. Each child is given a small oaktag card that is in a pocket on his or her desk. When I catch a student "doing something good" I give them credit by putting a hole punch in their card. After they get ten hole punches in their card, they get to choose a fun treat out of a "treasure chest" —erasers, rulers, etc. The kids really get excited about this, and once again the focus is on positive behaviors.

Maribeth Berliner, *Grade 3, Vermont*

---

Student discipline is something that an educator must work at and learn. Handling problems, for even the most experienced educator, is a challenge. No one technique or strategy works with every student. For me I use structured instruction time, firm voice control, fair discipline when needed, supervised independent time, cooperative learning teams, respectful comments and lots of sincere positive praise. I know when students feel good about themselves, know the expectations and understand that they are learning, I'll have few problems in my classroom.

Dorothy Hartson, *Grade 6, New Hampshire*

---

During the first week of school, my students and I discuss how our classroom will best run smoothly and effectively. The children suggest rules which I post on chart paper. We then narrow it down to the five most important ones which are posted in the classroom. A discussion of consequences follows with students' input regarding broken rules. This is a democratic approach with children taking an active part. It is important for teachers to be fair and consistent in carrying out the consequences.

Sharon Conley, *Grade 2, Maine*

---

It doesn't take a book of rules or serious consequences to control behavior. First you must develop a good relationship with your students. Let them know you care. Work on building their self-esteem every day. Tell them how special they are. Show them the importance of respect for teachers and fellow students. You give respect and in turn you will get respect.

Discuss your disappointment in their behavior one-on-one. Remember soft-spoken words speak the loudest.

Eileen Bruton, *Grade 5, Wisconsin*

---

Before the school year begins, a teacher should always decide on a discipline plan which includes well-defined rules and consequences. This plan should be taught to the students on the very first day. The students need to know from the very beginning that the teacher is in control, and exactly what is expected from them. Consistency is the key to making the plan work. A threat of action should never be made without a firm commitment to follow through. Lack of consistency can cause a total breakdown of discipline in any classroom.

Kathryn A. Smith, *Grade 2, South Carolina*

---

Be consistent—make rules that will achieve your expectations of classroom behavior. Write the rules in a positive manner, at the language level of your students. Establish reasonable consequences. Remember, it is your classroom, the rules are your rules and you are in charge. Now, the key is to be consistent, relentless, and fair.

Kathryn Maloy, *Grade 2, Florida*

---

Be positive, consistent, and respectful. The first day of school have the children discuss what rules are needed to function in the room. The consequences are also discussed. The rules are modeled and practiced daily for a couple of weeks. The children can see what the behavior looks and feels like. Do not barter with the child about a consequence.

Also, on the first day, the children are given a pocket to hold tokens. When a child is seen practicing good behavior the child is rewarded with a token. When the child receives ten tokens the child counts them to the teacher. The child trades the tokens for ten minutes of free time.

Mary Geren-Saggau, *Grade 1, Nebraska*

---

I believe that discipline is a positive, orderly, and motivated response on the part of students towards a teacher's management plan and instructional messages. To achieve a well-disciplined class, I feel there is no substitute for a teacher's sincere devotion in maintaining continuous efforts toward satisfying the needs of each student. A teacher's paperwork and planning cannot interfere with the necessity of being available to students at all times, and, therefore, cannot be completed while students are in class. This requires an energy that permeates the entire school day. Discipline will be a natural outcome for a teacher who is always well-prepared, has, and is consistent in carrying out, a positive and reinforcing management plan, and who enhances her instruction with enthusiastic, hands-on lessons.

Pauline McCollum, *Grade 2, Florida*

---

I found over the years that I can resolve most discipline problems by being fully prepared for class. On occasions when materials are not ready and students have to sit and wait, that's when the noise and fooling around begins. I plan carefully for every activity. I try to have all materials laid out and easy to use. I also plan alternative activities for faster students. They have options to use hands-on or quiet activities if they get done quickly. When teachers don't have options ready, many students who finish quickly will become behavior problems.

Mary Ann Gildroy, *Kindergarten, Montana*

---

Always go over the classroom rules the first day of school, and continue this process for about a week. Call on different students to state the rules, and have the class role-play the rules.

Velma Bell, *Grades 3 – 5, Learning Disabled, Georgia*

---

Classroom discipline begins with respect. Respect each child that crosses the threshold of your classroom. Children are very intuitive creatures. They will respond to your consideration and behavior toward them by rising to your expectations. High expectations of the entire class will also lead to a positive learning environment. Children always live up *or* down to the perceived expectations set for them. Expect appropriate behavior from all of your students.

Becky Ann Erickson-Danielson, *Grade 1, Minnesota*

---

The best advice I can give is to let the children know that you are in control of the classroom. There is a fine line between being teacher and buddy. It is hard to be both. Your top priority is as teacher, and you and your students must know that you are in control. Remember, idle minds and idle hands can be your biggest discipline problems.

Stacy Burns, *Grade 1, Mississippi*

---

As soon as possible in the school year, provide parents with positive feedback (postcards, notes, phone calls) about their child. This shows the parents that you feel their child is important and that you have something positive to say about their child. The parents are then willing to work with you if a problem comes up later in the year. Be fair to all the children in the room, allow them to tell you privately how they feel about your discipline policies. This information can help you evaluate your policies, and whether or not the policies are working for all your students.

Marcia Stirler, *Grades K – 3, Iowa*

---

Make your expectations clear.
Establish a simple record-keeping system for disciplinary incidents.

Note infractions for each student on a library card and reward specific acts of self-discipline with stickers.

Provide natural or logical consequences. ("If you don't turn in work, you'll have to give up recess to complete it.")

Provide incentives. For example, students who collect a specific number of stickers earn lunch with the teacher or a special snack.

Cay Spitzer, *Grade 6, Colorado*

---

*Assertive Discipline* by Lee and Marlene Cantor has worked miracles for me. I do have a right to do it my way and still teach the children.

Peggy Heneveld, *Grade 6, Alabama*

---

The key is *consistency*. To have good discipline, you must deal with each situation, consistently. Post room rules and consequences for breaking them. The first time a rule is broken the student is warned by writing their name on the board. Each subsequent infraction is followed by a check mark after their name. Consequences become more severe with repeated infractions. Refer to *Assertive Discipline* by Lee Cantor for more details.

John Myers, *Grade 4, Michigan*

---

On the first day of school, the teacher and students work together to come up with classroom rules. After listing them, post rules at the front of the room. The children should then tell the importance of each, understanding that together they can have classroom control with an effective learning environment. However, discussions need to address what will happen if one decides not to follow the rules and the teacher needs to follow through accordingly.

Joe F. Coleman, *Grade 4, Ohio*

---

There has been so much written about assertive discipline. I use it to a degree, but at times, one has to just be firm and say, "No!" Students want discipline and will abide by the rules you set down as long as you are consistent. I never give ultimatums because then I'm forced to comply. Once my students know where I stand, I have little problem getting and keeping their attention.

Ron Baysinger, *Grade 6, Indiana*

---

Discipline is the key to a successful and fulfilling teaching career. Without discipline, even the best of lessons and plans are lost. A teacher must be firm, fair, and consistent, and discipline with respect and kindness, not anger. The student needs to be aware of what you expect of him and what the consequences will be should he fall short of these expectations. Consequences should be reasonable and attainable (no idle threats) and you should always follow through with them. Once your students know what you expect and that you will hold them accountable for their actions, disciplining and teaching are easy.

Daria Ann Wood, *Kindergarten, Wyoming*

---

Establish a realistic and appropriate list of no more than four basic classroom rules. Rules should be global and deal with major issues which affect the learning climate for all students. Ignore minor offenses if they do not interfere or disrupt the learning environment of other students. Be consistent with students who violate basic rules. Reward and/or praise frequently those who consistently follow the rules.

Helen M. Martin, *Grade 4, Georgia*

---

Have a strategy prepared before school begins in the fall. Present this plan to the students on the first day of school. An example of this is to allow the students in a particular subject area to earn up to five points per class period. Points are earned for time on task, completed homework, etc. When thirty points have been earned, these convert into thirty minutes of game time.

Barbara J. Bell, *Grade 4, Pennsylvania*

---

Always try to deal with classroom discipline in a positive way. Be consistent and firm with students as you enforce the rules necessary to teach. Assertive discipline is a very effective model to follow. It is positive, and it puts the problem student back in the hands of the parent. Remember to always look for and dwell on the positive behavior from each student and conference privately with the student about bad behavior.

Penny Parnell, *Grade 3, Alabama*

---

You and your students together develop a set of acceptable behavioral rules the very first day of school. Establish consequences for breaking any of the rules. Display a classroom chart for each. Keep the list brief, four items, so they can be monitored easily. Reward good classroom citizens with verbal comments, write names on a reward chart, give stickers, etc. Be fair and consistent, yet kind and firm. Let your students understand you are here to teach and *no* student will keep you from doing your job.

Betty Robeson, *Grade 3, Maryland*

Each student is given a copy of the class rules and three "discipline cards" (index cards with the pupil's name written on them) to keep in an envelope in his/her desk all year. When a class rule is broken a card is collected and dated. If three cards are taken in one day, the student stays after school or misses recess. At the end of each week any student who hasn't lost a card is rewarded. These dated discipline cards are a good record of a student's behavior, and can be very helpful at conference time.

Wendy Romano, *Grade 3, New Jersey*

Establish a set of rules with the children as the school year begins. Children need to know the guidelines, and that they will be consistently enforced. During the first two weeks, I find I remind the children often about our classroom rules. Some children need to challenge these rules to ensure the rules will be enforced. After this initial period, very few reminders are necessary, as the children accept the standards for behavior.

Shirley H. Goodemote, *Grade 1, New York*

Problems with classroom discipline just before a holiday can be eliminated by providing a "real" learning activity to fit the holiday. At Christmas read *Millie's Book,* by Barbara Bush, and then have the students write letters to Millie and color ornaments for her Christmas tree. When the letter to Millie is finished, students can write to the Prez and Bar of *Millie's Book.* This fun activity keeps students happily focused for several days. Writing a letter is the teacher's objective; fun is the students' objective.

Bonus: All students will receive a picture of Millie and Barbara Bush.

Shirley Spear, *Grades 4 – 6, Language Arts, Texas*

Do not be afraid to stop your lesson and start again if discipline disrupts learning. Take a deep breath, stay calm and stop all activities. When the students are silent and still, and you have everyone's attention, state what needs to be done and how. Slowly begin again. Your firm discipline is more important than rushing through a lesson just to meet your deadline.

Discuss, list, and post your expectations for the class in positive examples such as: be proud of yourself; be honest; be helpful; be a good student; be cooperative; be compassionate; be calm.

Review rules *constantly.*

Janet Mar Dong, *Grade 5, California*

At the beginning of each school year, the teacher is to convey to the students what type of conduct/behavior is acceptable/unacceptable. When handling discipline problems, the teacher should always remember to be consistent. Each student should know who is in control during the first few days of school. Children should not only understand the discipline plan, but parents should understand it as well. Once rules have been established, there should be no deviation from the plan.

Linda N. Stevens, *Grade 1, Mississippi*

---

Have a written discipline plan posted in the classroom. Include expected student behavior, consequences for noncompliance and rewards for compliance. Go over the behavior/discipline plan in detail on the very first day of school. Keep the expectations as positive statements, i.e., "Keep hands and feet to yourself," instead of "No hitting or kicking."

Christine Shepard, *Grade 2, Utah*

---

Respect your students. Discipline will flow with the daily routine if students feel respected. A student should never be belittled or made to feel unvalued. Sarcasm does not have a place in student discipline. Instances where a student needs to be corrected or counseled should always be a private situation.

Lesa Carroll, *Grade 5, Missouri*

---

To create a classroom environment with basically no behavior problems, the children have to *want* to behave. I have done this by creating an environment of high control and high support. This means that I constantly reward my class in many different ways when I see positive behavior, or the kind of behavior that I want to see in my classroom. I use verbal praise, class rewards with a star chart, and individual rewards through daily behavior folders. High control means that I do not allow any misbehavior at all. I do not raise my voice or interrupt my lessons to discipline, just put a name on the board. It works!

Sharon Kelly, *Grade 3, New Jersey*

---

The first week of school, have the kids come up with two or three rules by brainstorming a list. They'll want to list the do's and don'ts to add to the list. Combine rules such as:

- We/I will not hurt others.
- We/I will not mess with other's stuff.
- We/I will be good "teachers."

After the rules have been established and *signed* by each student, practice for full understanding. Infractions should be dealt with by the teacher and the individuals to be disciplined, never as a whole classroom.

Leslie More, *Grade 2, Washington*

---

Students love parameters and need to know the limit. I have a chart of classroom rules and consequences for breaking the rules. This helps me to deal with similar situations in similar ways. Grouping students in pods (or tables) and having a table captain on a weekly rotating basis helps students become responsible for their own behavior. Tell students when you demand quiet listening time, and allow them some freedom during work time. Encourage cooperative learning and allow quiet talking as long as it focuses on student work. You don't have to be in total control! Give students opportunities to make their own decisions, and learn from their mistakes.

Barb Parmenter, *Grade 4, Michigan*

## CHAPTER HIGHLIGHTS

When listing your rules, let the students help. Be sure all student expectations are written in a positive manner. A teacher must be consistent and fair, and follow through with discipline consequences. Be sure parents are aware of your classroom rules and student expectations.

## CHAPTER 3

# Motivating Students

IT is pretty well accepted by all teachers, no matter what the level of their classroom, that if students are going to realize academic achievement, they have got to be motivated. A very few students in any given classroom will seem to always be motivated to learn. A modicum of the class will begin the year with obvious motivation, but within a relatively short time, the overall atmosphere of motivation will probably drop appreciably, perhaps measurably, if you as the teacher don't do something to meet the condition head-on.

Just like the other problems first-year or early-career teachers face, this business of motivating students has no one strategy that will serve all teachers, all classrooms, or all students. Think about it. What motivates you? How about your brother or sister? How about your friends? Your peers? Chances are very good that what motivates the people around you varies significantly from one to the other. So why should you expect it to be any different in your classroom? You shouldn't. There are, however, a couple of basic tenets endorsed by most teachers regarding student motivation. First, you have got to set the mood or create the atmosphere which encourages student motivation, and that is accomplished very simply by *your* motivation. There's an old Texas saying that goes: "You can't start a fire with a wet match." The classroom correlate is that you won't ignite your students if you don't have any spark. Keep that in mind.

A couple of other student motivation basics which seem to be a part of the operational repertoire of successful classroom teachers are the generic entities of individual student achievement and positive reinforcement. There are many, many specific applications or strategies by which each of these basics can be realized. The following are successful techniques which have been field tested and endorsed by exemplary veteran teachers.

Be motivated yourself! Don't prescribe work, lessons, schedules or anything else that you are not excited about. If you spend the time to keep yourself jazzed after teaching the same thing for many years, for sure the class will be motivated. The quick and easy, tried and true will burn you and the kids out quickly. Get into a learning feeding frenzy!

Leslie More, *Grade 2, Washington*

---

I find that giving rewards for a job well done is an excellent motivation for students. Therefore, students who finish work early are allowed free activity time after weekly tests are completed. This gives them an opportunity to enjoy their favorite class-related work. This also motivates other students.

Linda N. Stevens, *Grade 1, Mississippi*

---

Provide models of what is expected. Show students step-by-step what you expect them to do, then provide time to go through that exercise.

Immediately, have each child tape up his work on the board. (Give them a small piece of tape to do it themselves.) *Every* child *must* show his/her work. As you do this frequently, you will see the quality of work improve each time. It's teamwork, and each member improves. It is not always necessary to mark or grade *every* set of papers.

Janet Mar Dong, *Grade 5, California*

---

It is true that students respond to praise and positive reinforcement.

A good way to motivate students is to have a jar in the classroom. When an individual, or the class as a whole, does something well, add popcorn to the jar. Popcorn can be earned for using good manners, helping others, quiet dismissal, being on task, and so on. When the jar is full, pop the corn and have a snack.

Barbara J. Bell, *Grade 4, Pennsylvania*

---

Students are information rich and experience poor. Our schools were set up for the experience-rich, knowledge-poor student of the 1800s. Give the students activities and the rest will take care of itself. Burn your basal and worksheets.

William Andrews, *Elementary, Massachusetts*

---

It is my feeling that children are naturally motivated, and that as the adults in their lives, it is our responsibility to nurture and enhance their curiosity, and help them find success in their efforts so they will continue to be motivated even after they leave our classrooms. I like to design my program around their interests, which they see as ''validating'' their experiences. It is the artful teachers who can integrate their objectives and curriculum into content that sparks the attention of their children. To keep the sparkle in the children's eyes, and keep them motivated to learn, it is important to *know the children*. Further, as a primary role model in their lives, model enthusiasm.

Maribeth Berliner, *Grade 3, Vermont*

---

All motivation techniques should begin with the building of the students' self-esteem. All students, whether high achievers or low achievers, have a need to believe that they can and will succeed. Providing students with successful learning experiences clearly motivates them to attempt further challenges.

A teacher must also demonstrate to the students that learning is fun. Materials should be stimulating and inviting. The teacher's presentation of the curriculum and the materials should be equally as exciting. Excitement in learning is contagious. A teacher can spread it by being a strong role model for the students to follow.

Kathryn A. Smith, *Grade 2, South Carolina*

---

A monthly homework report card is given to encourage students to complete homework, and to communicate pupil progress to parents. A rating of excellent, satisfactory, or unsatisfactory is noted, and incomplete homework assignments are listed. Students achieving an excellent rating are rewarded with a free homework coupon good for one assignment in the following month.

Wendy Romano, *Grade 3, New Jersey*

---

Always try to accentuate the positive. When grading a paper, write the correct number of problems over the total number of questions (i.e., 7/10 instead of −3). Also, highlight correct answers with a yellow marker instead of marking incorrect answers wrong with your red pen. Gold paint pens are great for recognizing perfect work. These are 14K gold papers.

Personalizing as much as possible is another great tool for motivating students. We have been concentrating on story problems in math. The students can hardly wait to get the problems each day because we include the students' names in as many as possible.

Instead of marking an X on a graph or coloring in a box, xerox students' school pictures for graphing purposes. Students love to put their picture beside their choice on a graph.

Charlotte Hargett, *Grade 1, Mississippi*

Kindergarten children are easy to motivate. I feel it begins with the enthusiasm shown by the teacher. If I am excited about a unit, the children are more interested in it. Candy M&M's, stickers, happy faces, special prizes for contests (everyone wins), praises and more praises, smiles, and hugs – all work great. Show a sincere special interest in the child. Let each child know what you expect from him.

M. Faye McDonald, *Kindergarten, Mississippi*

Motivating students is a challenge. Plan lessons well. You must attempt to understand your students' needs, wants and way of learning, and to establish a sincere rapport. Centers, manipulatives, cooperative learning, projects, story-telling, and a large inventory of fiction/nonfiction books spanning several age levels and many subject areas, will motivate children to learn new things. A sense of humor and interesting lessons are more effective than textbook reading – children have difficulty relating textbook information to everyday life. You'll know when children become disinterested, and you must be willing/able to change gears and use a different strategy to reclaim their interest.

*Anonymous, Elementary, Rhode Island*

Praise children for positive actions. Choose the traits that you want to encourage and compliment students who exhibit these characteristics. Tell your students how wonderful they are in many ways (i.e., "I like the way you . . ."; "You are so clever to . . ."; "Wow, I'm surprised you already know how to . . ."). Children improve faster if they are recognized for even small achievements.

Marilyn K. Emmons, *Grade 1, Montana*

A lot of times older students (L.D. students!) come to my classroom to read books they have authored, plays they have written, or puppet shows they have written and produced. After listening to our guest reader we take a few minutes to draw our favorite part of the book, play, or show. We then combine the individual pictures into a special thank-you packet, along with a note to let them know what a wonderful job they did. These packets become prized possessions of the students that are shared with their teachers, fellow classmates, and parents. More visits are usually made – and requested!

Barbara Herman, *Grade 1, Transition, Oklahoma*

Provide an I-can-do-it atmosphere in your classroom. Preempt each new challenge with an attitude that the whole class will succeed. Encourage students to discuss assignments in pairs or groups.

Barbara Baker, *Grades K – 6, Hawaii*

---

This is probably one of the toughest problems to overcome. At the beginning of the year, with a new teacher, everyone is excited and dives into the challenges you provide. After the newness wears off, those students not easily motivated begin to surface. I take the direct approach. I sit down with the students one-to-one, and I ask them what interests they have, what I could do to help them get the most out of school, what they think they could do themselves, and what we could do together. I try to help them see the value in what we're doing and how it pertains to them personally. If they perceive an activity as worthless, they aren't going to put any effort into it.

Nancy Mahloch, *Grade 6, Nebraska*

---

I find the key to motivating students is to keep them all actively involved. I use their names in problems, tell stories that they can relate to, such as stories about my family and my own childhood. I begin every lesson by relating the current lesson to some prior knowledge and I allow for every child to feel success.

Sharon Kelly, *Grade 3, New Jersey*

---

Let your students know you expect their best effort on all of their work. An art project is just as important as a math test. Praise the students for their efforts as often as possible.

Christine Shepard, *Grade 2, Utah*

---

Gearing instruction to the level of each student with creative and relevant activities is the most effective motivational technique. When students are asked to complete tasks too difficult, or that are above their individual performance level, they are being set up for failure. I have found when students can move with success from an easier to a more difficult task, it is easier to motivate them in future activities.

Helen M. Martin, *Grade 4, Georgia*

---

The teacher needs to find the interests of the children and then relate those to the studies. One needs to find something that turns the students on.

We have to keep the students' interest or we may fail as educators. The children need to see that we are working with them, that we believe they can succeed. One needs to understand that children are individuals and may need to try one-on-one contacts.

*Joe F. Coleman, Grade 4, Ohio*

---

One key to motivating students is to be enthusiastic about what you are doing. Don't expect students to be excited about a lesson if you're not. Also look to use students' interest areas in making applications to your lessons. Stay away from repetitious busywork. Using visual aids and/or hands-on manipulatives will help make learning more effective. Finally, if students can see practical applications to what they're learning, this will help motivate them.

*John Myers, Grade 4, Michigan*

---

Once a week we have a special day — Terrific Thursday, Fabulous Friday, etc. Students who complete their assignments and haven't gotten in trouble earn the privilege of participating. Other students follow a regular routine (no punishment) while the ones who earned it are rewarded with a day of games that teach the skill. Of course they say they have no work, only fun and games.

*Peggy Heneveld, Grade 6, Alabama*

---

"Success breeds success" is the key to motivating students. A teacher's responsibility is to plan meaningful, thought-provoking lessons in which students can achieve. Each small triumph gives a child a sense of accomplishment which will in turn become a desire to learn more. The confidence built in small steps motivates students to achieve and gives the child confidence needed to try new tasks.

*Becky Ann Erickson-Danielson, Grade 1, Minnesota*

---

A classroom that is vibrant with color, and displays student work, has lessons that are well-planned and that frequently involve student participation and hands-on materials, contributes to a motivated student. My students frequently ask where the day has gone, because they have been kept involved and busy (not involved in busywork) throughout the day. I believe that a motivated student is a student who receives positive reinforcements for success and/or efforts·to succeed. Therefore I feel that specific goals and rewards must be identified and frequently forthcoming. It is important to set both daily and weekly goals and rewards, and to set both individual and group goals and rewards.

Pauline McCollum, *Grade 2, Florida*

---

Be interested yourself, use visuals, manipulatives, items kids can touch, feel, or move. Be animated. Always have a motivation; introduce your lessons in a manner which will grab the attention of your students.

Kathryn Maloy, *Grade 2, Florida*

---

The best way to motivate students is to show individual interest. You must know your students' interests and backgrounds. Once you know whether a student responds to praise, or a sticker, or just a smiley face, you will be able to encourage growth in many areas.

To motivate students, you must be very organized as a teacher. You'll need to keep track of student work over a period of time, and then you need to make note of change and improvement. As a student realizes you are going to care what he does, he will be motivated to meet your expectations.

Finally, a little bribery of stickers, treats, free recess time, etc., will serve as an initial motivator—leading, hopefully, to a development of self-motivation in regard to doing a good job just to learn something.

Sharron Baird, *Grades 2 – 4, Oregon*

---

You really need to know your students. Find out from them what their parents do for a living. Show an interest in each child by asking about hobbies, pets, favorite sports, or after-school activities. Be available to students when they first come into your room each morning. It's best if you can greet them at the door. Many times they will want to speak with you first thing in the morning. Take some time to listen, and in doing so you may have a much better day, because you will better understand what has happened in their lives since they left you the day before.

Frankie C. Bly, *Grade 5, Minnesota*

---

I use positive reinforcement in every way, from having good achievement charts to giving handshakes. It is important to give positive reinforcement *immediately* after the proper behavior is achieved.

You need to be excited about the areas of the curriculum you teach. I do a lot of hands-on activities to get my children involved. Once they have a taste of success, they are willing to try harder and want to learn more. It's a step-by-step process—step . . . success . . . step . . . etc.

Debra L. Mann, *Grades K – 5, Special Education, Massachusetts*

The children are told their classroom is an "I-can" classroom and they are "I-can" children. Every morning they are asked, "How are you?" The children respond, "Super-super great and getting better!" On each child's desk is a thematic cut-out which says "I can." The children sing an "I-can" song and recite poems that affirm the "I-can" feeling. Using this power thinking tool, the children believe they can do any task.

Mary Geren-Saggau, *Grade 1, Nebraska*

Students are best motivated through a positive attitude, which comes with a positive atmosphere. Students need to feel good about themselves. Setting goals is important in the classroom, but you also need parent cooperation to really be successful. A smile, a pat on the shoulder, and a positive statement or compliment go a long way.

Pam Smith/Terry Carter, *Grade 6, Illinois*

To motivate students, motivate yourself, seek out new ideas, a variety of experiences for students to learn from, and be sure topics are timely, theme oriented, and/or age appropriate. If you enjoy the topic and are knowledgeable, your students will be motivated. Present the lesson, follow through with the lesson, and, when completed, evaluate the lesson.

Dorothy Hartson, *Grade 6, New Hampshire*

"I can't believe it. This chest was sitting right outside my door this morning. It's filled with money. Hurry! Help me count it!" The best motivator is excitement. All teachers have the ability to bring it into the classroom. Keep your students guessing. What's he/she going to do next? Keep a positive classroom atmosphere. It's Wednesday, the science experiment didn't work, and Mary has John's pencil. Let your students know today was okay, but wait until you see tomorrow. It will be great! "We're going to turn a liquid into a solid and then. . . ."

Eileen Bruton, *Grade 5, Wisconsin*

First, find out what interests the students, and from this list develop a reward check list. Allow the students to use the list when they have done something special.

Velma Bell, *Grades 3 – 5, Learning Disabled, Georgia*

Sometimes I feel even if I stood on my head I couldn't get a few students' attention. The best method I know to motivate students is to veer from the text from time to time without their prior knowledge. That way, they never know what to expect and will be more interested to go to different sources on their own. I've found that the use of resource materials is an excellent source of motivation.

Ron Baysinger, *Grade 6, Indiana*

---

Know your students. Know their likes and dislikes, their backgrounds, their interests. *Talk* to your students. Talk about their interests, their likes and dislikes, their feelings. Use this information to encourage them to improve themselves. *Praise* your students; better yet, teach your students how to praise themselves, how to work for personal gratification. *Encourage* your students. Build their confidence in themselves. *Challenge* them. Give them tasks, not necessarily academically related ones, at which they can be successful.

Daria Ann Wood, *Kindergarten, Wyoming*

---

Students are more easily motivated when you give them choices and you focus on things that interest them. Let students choose books they want to read and then you can put in required reading. Give students opportunities to be a part of decision making in the classroom rather than always dictating what to read. We as teachers must stay read up and keep interesting materials, books, etc. in the classroom. Remember we must model what we want students to be.

Penny Parnell, *Grade 3, Alabama*

## CHAPTER HIGHLIGHTS

A teacher must be enthusiastic about the subject matter and lesson being taught. Pass this on to your students and let them experience success. Present attainable challenges to your students and then reward them. Students will learn best when directly involved.

## CHAPTER 4

# Classroom Management

CLASSROOM management is another challenge, if not a problem, that pervades the ranks of all classroom teachers. It is not a concern that is limited only to first-year or entry-level teachers. It is an ongoing concern, to some degree, no matter how long one has been a teacher. However, it is during that first year that a teacher seems to be bombarded with an overwhelming need to organize things—student issues, administrative issues, instructional issues, classroom issues, and personal issues. At the very best, the condition is stressful, if not somewhat frightening. Add to all this the fact that classroom management will definitely have an effect on classroom discipline and student motivation. This impacting characteristic of classroom management is why it is so important that this particular problem of early career teachers be taken very seriously, and focused on as soon as possible.

Veteran teachers know that having control of this business of classroom management does not just happen. First of all, it does take extra time on the part of the teacher. It is a commitment. It requires good planning. This is also a part of a teacher's life in which flexibility is a good quality—if you have it. In fact, this is one area in which you may wish to entertain a paradigm shift.

Many successful teachers recommend involving the students in classroom management. Not only do students provide a larger work force, but they find that the involvement is contagious, and becomes personalized to the extent that the effect spills over into the students' classroom parcels.

Just as we all manage our personal settings in life somewhat differently from one another, so it is with teachers. It is ludicrous to suggest teachers should all manage a classroom in precisely the same manner. The only constant in classroom management is the differences between individuals. This is where you can probably do your thing—as long as it proves to be effective for you.

Consider the variety of approaches that the veteran exemplary teachers have suggested, do a little mix and match, and good luck!

---

Number file folders from one to thirty-one (days in a month). Put these file folders in your desk drawer. When you make out your lesson plan put the materials needed for that day in the correct file folder. This way you or a sub knows right where all the materials are for the next day. Have a file holder in the room where the students can put their papers under the right subject. This will save you a lot of time when you go to correct the papers.

Marcia Stirler, *Grades K, 3, Iowa*

---

A well-organized school day is very important for our students. Teachers must be willing to give the extra time to be well-prepared. Integrated units and centers need to be prepared in advance with careful attention to the needs of all students. Children's interests and ability may take our teaching of a unit in a different direction, and we, as teachers, must be prepared to take their lead. Flexibility is a key word in classroom management.

Sharon Conley, *Grade 2, Maine*

---

When the children arrive in the morning, they unpack and look at the get ready directions listed on the chalkboard. At this time they place homework papers in designated areas, sharpen pencils, head their spiral notebooks in all subject areas, and begin an enrichment assignment. The latter can be a practice curriculum worksheet, a journal writing, or a current event activity. This is worked upon throughout the day any time an assignment is completed. When I want the attention of the entire class, I ring a small bell rather than raise my voice.

Betty Robeson, *Grade 3, Maryland*

---

Try to plan a full week in advance. Even if plans change, you will still be prepared. Design a system of reminders that will alert you as you scan your plans through the week that certain materials are needed for particular classes. Try to minimize the transition time between lessons. If you are organized and ready to go, your students will be as well.

Lesa Carroll, *Grade 5, Missouri*

---

Classroom management can be made easier by good organizational skills. Group students in pods of four to six. This gives you more classroom open space, papers and materials can be passed out quicker, it allows for small group control making students responsible for group behavior, and it allows for easy cooperative learning groups. Use a *checklist* of activities on the board so students have a visual reminder of things to be done. Use individual file folders for student papers. Let students file graded papers during free time. Pass out folders on Friday—it saves so much time. Establish classroom rules and consequences—chart them and post them—students then know the limits and it helps keep teachers fair.

Barb Parmenter, Grade 4, Michigan

---

I always have the next day's assignment ready the day before I leave for home. It is laid out and organized by order of presentation. I plan the week's work in advance, running off materials, etc. for that following week. In my room I organize things with colorfully labeled boxes so students will know where work goes (finished, homework, etc.). I have a mailbox for homework and letters going home. Students keep a weekly assignment sheet to organize their subjects. This teaches them organization, and enables them to know what is finished, and what is not done.

Karen Morgan, Grade 6, Idaho

---

Make your classroom an inviting, nonthreatening place to be. It sets the mood as the children enter each day. Desk arrangements can be as varied as the goals for your class. Include children's work in classroom decorations. Be prepared. Choose applicable objectives on an appropriate level. Teach the objective in a variety of ways using varied tools of teaching. Encourage student participation. Be brave—try new ideas. Be positive—children will perform better if they respect you and themselves. Be flexible—always have a Plan B.

Anonymous, Grade 4, Utah

---

Be sure to check your state's time allotment requirements for each subject. Ask your principal for past schedule examples and for advice. Label a folder for each day and put that day's worksheets and other project plans into it.

Faye Karna, Kindergarten, North Dakota

---

I like to come in early enough to organize my day's work on my desk. At the end of the day, one's mind is at a point of rest. My mind starts thinking about the day even before I arrive at school. I have to know what I am to do all through the day. I always have more laid out than can be done. This is an essential element of each day.

Stacy Burns, *Grade 1, Mississippi*

---

It is very important that a teacher be organized, and the classroom be well-organized. There must be a place to return all learning materials after use. Remember to use students to help keep the room organized. No two school years will ever be exactly alike. Different students bring diverse needs into your classroom each academic year. Involve students in grading class work. Give immediate feedback from test and classroom work by allowing students to grade their papers as you call out the answer.

Penny Parnell, *Grade 3, Alabama*

---

Be prepared and organized. Have your materials ready and at your fingertips. Take the time to plan out your year, your month, and then your week and have the materials ready for a week in advance. This alleviates unplanned vacation. Set up a file system in your room and have a place to put each day's papers and supplies. Set up three files right away, and then take the time to file items as soon as you are through with them. I have a file for bulletin boards and classroom displays, one for projects and art ideas, and one for thematic units, possibly one for storybook units. The key is to be prepared. If you know what you are going to do and have the materials ready, the entire day will go smoothly.

Daria Ann Wood, *Kindergarten, Wyoming*

---

Whenever a new student enters, always pair the child with a student mentor, one who has demonstrated a knowledge of all the class rules and schedules. Students seem to learn a lot from each other, plus this skill also increases social interaction.

Velma Bell, *Grades 3 – 5, Learning Disabled, Georgia*

---

Organize your worksheets into folders labeled with the days of the week. Take out the appropriate folder for that day. A posterboard pot and spoon hang on my classroom wall. In the pot are all the students' names. Each day a new name is picked and put on the spoon (pot luck). That student is responsible for the classroom duties. This replaces twenty hands going up for each task. Trays labeled ''Today's Work'' and ''Late Work'' help organize my correcting. Cardboard shoe organizers substitute as mailboxes for returned work that goes home on Friday.

Eileen Bruton, *Grade 5, Wisconsin*

An effective school day begins long before 8 A.M. All materials must be ready to be used. All lessons must be prepared with special attention to the group who will be using them. Your work begins as soon as you know what area you must cover. How you cover that material and meet the needs of your students determines your effectiveness.

A good teacher covers the curriculum, but even more, he/she *causes* learning to occur. Getting students interested and enthused about school can mean you are effective. The classroom environment can also greatly affect student learning. The arrangement of furniture, providing an area for reading or learning activities, and highlighting work or learning, are means of gauging teacher attention to detail and student needs.

Sharron Baird, *Grades 2 – 4, Oregon*

Be flexible. If something does not work for you, change it. Most schools have schedules for special classes, lunch, etc. that you must work around. Plan your schedule to fit the needs of your children. Move children's seats frequently in order to meet the specific needs of both student and teacher.

Kathryn Maloy, *Grade 2, Florida*

Each of my students has a work folder which contains an awards collection sheet attached to one side, and a grading sheet attached to the other. Students earn stickers for completing daily work. Collecting stickers earns both daily and semester treats. I avoid the loss of instructional time by eliminating the passing out of papers throughout the day. All work for the day is placed on a table. As each student arrives in the morning, he takes one of each assignment and places it in his folder. The students turn in their folders with completed work at the end of each day. I return the folders the following morning with papers graded and grades recorded. I also have an activity wheel. This is rotated daily, and indicates whether a student should visit the library, use learning centers, or enjoy a free choice day. Using this wheel ensures equal distribution of activities and releases additional time when I am free to assist students with their assignments.

Pauline McCollum, *Grade 2, Florida*

Organize the classroom learning environment to be as efficient and effective as possible. Arrange learning centers which correspond with the units being taught throughout the classroom. When children finish their work, they know where to go and what to do. One of the most important areas to incorporate into classroom design is a comfortable reading area with a multitude of books for the children to enjoy. The teacher's desk should be low profile. The students are the center of attention, not the teacher.

Arranging the school day is a delightful challenge. Integrating all content areas helps to arrange the class work and motivate students at the same time. Take into account the individual learning styles of all students when planning the lessons for the week.

Becky Ann Erickson-Danielson, *Grade 1, Minnesota*

---

I prefer to write lesson plans one week in advance. I write them the previous Friday and on Monday morning I get assignments run off and gather materials for the week. I prefer a neat, colorful, uncluttered classroom. I find that desks arranged in rows allows students to concentrate better. Also, if you can develop a *routine* preparation schedule as well as a *routine* daily class schedule you will get the most out of your valuable time.

*Anonymous, Grade 4, Michigan*

---

With more and more emphasis on how the teacher needs to be accountable, and with efforts being focused on nationwide student testing, it is more important than ever to have an effective school day — five days a week. The students need to be kept on task. Let the children know the importance of being involved the whole day. If one does not use the time wisely, it could result in a great loss of the students' education at the end of the week.

Joe F. Coleman, *Grade 4, Ohio*

---

Space is a difficult problem to resolve. The successful classroom teacher incorporates large-group, small-group, and individual classroom experiences. Rotating individuals and pairing students to complete tasks is a meaningful and successful method for effective learning for both the handicapped and the gifted student. Provide students with the opportunity to move to different work spaces for variety. Change the arrangement of student desks from rows to clusters depending on the activity.

Helen M. Martin, *Grade 4, Georgia*

---

Have a schedule written on the chalkboard every day. List the students' assignments in the order that you want them finished. I use four trays to collect assignments from the schedule. Each assignment listed in the schedule has a tray number after it. We have direct instruction lessons throughout the day, but whenever that work is finished, students know that they should go back to their schedule.

Christine Shepard, *Grade 2, Utah*

---

Be as organized as possible. I use a lot of small labeled and color-coded baskets. I have my class set up for the students. There is a reading corner, arts and crafts supply area, paper area, etc. The important thing is that materials are stored where the students can get them and return them. Seating arrangements are kept very flexible. I change the seating every month using a wide range of floor plans. I basically always keep a 9′ × 12′ carpet piece in the center for whole group instruction and story times, and I have the desks arranged in groups of four or five, in pairs, or in a modified horseshoe.

Sharon Kelly, *Grade 3, New Jersey*

---

This is one area that my undergraduate courses didn't really prepare me for. I was lucky to have been assigned a "buddy" teacher my first year. That helped me out a great deal. She showed me class schedules from other years so I could get some ideas. My advice would be to find a "buddy" in your building. I have since learned for myself that it is easier to first schedule those blocks of time where your students have to be out of the classroom, such as music, physical education, library, computer, recess, lunch, etc. From there I fill in the subjects that I am responsible for. I have found that afternoons seem to be the time of day when students are a little harder to motivate, so I schedule those classes that are more active. Science is one example, because students can be busy working on experiments, etc. As for the physical arrangement of the room, I change it around at least once a month, sometimes more. I let the students help in the decision making. It helps them feel more in control of their surroundings.

Nancy Mahloch, *Grade 6, Nebraska*

---

Begin with a pleasant classroom physical order and environment. Children will have an easier time organizing their desks, assignments, and thoughts if the classroom is clean, orderly, and pleasant. Involve students in the maintenance of an orderly classroom and provide such materials as containers for centerpieces, bookends for classroom library, etc.

Barbara Baker, *Grades K – 6, Hawaii*

---

Here is how I handle first-of-the-morning lunch count, personal stories or experiences, money collecting, etc. At this age everyone has something special to share with the teacher *first* thing in the morning. So as soon as the bell rings to come in and start our day, my children either get a piece of paper or their journal and draw, color, and/or write about their special story or event. While they are engrossed in their personal work, I am able to take care of the lunch count or any other morning business. Afterwards, we share our stories and each student has mine and everyone else's undivided attention to hear their interesting stories.

Barbara Herman, *Grade 1, Oklahoma*

---

A schedule will help to organize you and let your class know what's expected every day. Allow enough time for your subjects, and recognize that sometimes you will exceed this limit, and sometimes you'll need extra activities planned to fill out the remaining time. Vary your activities within subjects to keep up interest levels. Organize your classroom with extra centers, so that children who finish early can have extra fun and practice on current skills.

Marilyn K. Emmons, *Grade 1, Montana*

---

An organized classroom maximizes teaching time by reducing the amount of time spent on routine tasks. Try these helpful hints: (1) organize teaching materials, maintain current files for each concept within a discipline; (2) establish monthly activity folders, file ideas for holiday projects or activities; (3) devise a simple method for distributing and collecting student work, select a daily student leader and set out labeled folders for completed work; (4) send home students' work once a week, use milk cartons as mailboxes, have students file their corrected papers, at week's end add notices, etc.; (5) establish several classroom centers where students can go when seat work is completed (a heavy plastic wading pool lined with blankets and pillows makes a great reading center).

*Anonymous, Elementary, Rhode Island*

---

When arranging the classroom, have good walkways so children can move about easily. Arrange centers so quiet centers are away from noisy ones. For instance, don't have the book center near the block center. Make thorough lesson plans to carry you through the whole day. I like to plan more than I will need and keep extras for emergencies. At the end of the day, set up your room for the next day so that everything will be ready when you walk in the classroom the next morning.

M. Faye McDonald, *Kindergarten, Mississippi*

We begin each day with whole group time. Activities from all skill areas are covered during this time. We then divide into five groups for table activities. I put all materials needed for each activity in a large plastic ice cream bucket labeled Activity #1, Activity #2, etc. When table time begins, those students who have been grouped together on Monday proceed to the next table from the one they worked at the previous school day. Oral instructions are given for each activity, and captains for each group are appointed on Monday. This saves time. Therefore, once the bucket of materials is distributed, the activity is ready to begin. Students may not leave the table until time for all students to return to whole group time.

Charlotte L. Hargett, *Grade 1, Mississippi*

A seat work rack is used containing a folder for each subject. A class chart listing pupils and assignments is in each folder. Pupils file work in appropriate folders as work is completed. A dependable student checks off names of students who have turned in work. At a glance the teacher knows who has or hasn't completed the work. Charts are also helpful for keeping track of homework, signed tests, and incoming projects.

Wendy Romano, *Grade 3, New Jersey*

Well-prepared, long-range and short-range plans are essential. Long-range plans should be made prior to the beginning of the school year. They provide a time frame to help assure coverage of all key elements of the curriculum. Short-range plans are crucial for the teacher of a smoothly run, organized classroom. A full week's plans should be completed and ready to follow on Monday of each week. Along with these plans, materials should be in the classroom ready for use. The materials should be separated and labeled to indicate the days to be used. Teachers who plan well have minimal loss of instructional time and fewer subject transition problems.

Kathryn A. Smith, *Grade 2, South Carolina*

Many times a student's desk isn't large enough to hold all of the materials needed during the course of the day. Time is lost while children search for lost items. A maintenance man from my district made a wooden storage unit for my classroom. Now each student has a compartment or mailbox. Each opening is large enough to store paper, folders, scissors, rulers, band music, and small art projects.

Barbara J. Bell, *Grade 4, Pennsylvania*

Provide a happy, warm, clean, and organized learning environment.

All materials should be kept in a sturdy container on a shelf at the children's level, i.e., homework box, scissors box, pencils, etc. Decorate the room with students' work in an appealing matter. Use color backings. Help students become proud of their work. Assign *each* student a task to maintain and organize in five minutes; thirty tasks can be completed to maintain a clean uncluttered environment. Initially the teacher must show what she/he *expects* at each task.

Janet Mar Dong, *Grade 5, California*

## CHAPTER HIGHLIGHTS

At the elementary level, when you have the class all day, being prepared helps make an easy transition from subject to subject. Use of charts, file folders, or containers to keep activities grouped will allow the students accessibility to their projects. Use your time wisely.

## CHAPTER 5

# Insufficient Classroom Materials and Supplies

AS you remember your immediate, recent, and distant past, you will probably agree that you almost always need, or at least want, something you don't have. So it goes for teachers. But their situation rings with a bit more urgency, since their declared insufficiencies are usually based on the instructional needs of their classroom population.

You will be a very lucky teacher if you never have to concern yourself with this problem. Most teachers will have to deal with this—usually on a continual basis. As long as we see this as almost a given in the life of a teacher, we may as well make the most of it—have some fun with it. Once you get into the specifics of ways to minimize, if not eliminate this condition, you will develop a great deal of confidence. First, you will have to accept a few simple postures:

- Be creative and imaginative.
- Improvise—use an item for something other than what it was originally intended for.
- Perceive your entire community as a resource base.

With the right attitude and an open mind, this problem of insufficient classroom materials and supplies won't register on your stress side. Start your own personalized solutions primer using the ideas that follow.

---

Don't go for glossy and cute. Start the year off with a unit on the environment and recycling, reducing and reusing. Enlist the help of parents to send in no-longer-needed items. Send home a monthly list of items you'll be needing in art, science, language arts, etc. Many businesses can help in this area as well. Help the children feel proud and responsible for bringing in materials as well as using and reusing materials. Make sure you value their efforts by practicing what you preach. Cover bulletin boards in newspaper or material scraps, have a large scrap paper box, make room dividers out of pop can plastic carriers, and display books on rain gutters. Your room does not need to look like a junk yard — just like you care.

Leslie More, *Grade 2, Washington*

Learning to be creative has helped me a great deal. Making things for many class-related activities is very inexpensive. I've also been able to substitute supplies and materials. I find that allowing students to write work from the board cuts down on having to make copies. This places the expense of paper on the parents. This also gives first-grade students a good opportunity to develop writing skills. Local libraries can also be beneficial in obtaining skill-related activities.

Linda N. Stevens, *Grade 1, Mississippi*

Businesses and parents can be helpful. Go to the Lions or Rotary Club. Tell them ''I need twenty-four beakers.'' If you can show this to the community, they will respond.

William Andrews, *Elementary, Massachusetts*

The problem of having insufficient classroom materials and supplies can be solved simply — be creative! Many of the things that classrooms need can be gathered from tag sales and, my favorite, the homes of your students. I send home a materials list with all students at the beginning of the year, with things listed that can be used for all the curriculum areas. Why not recycle those old, unused things, and at the same time supplement your child's education? I have found it possible to supply full science units and math stations through the combined efforts of home and school. My big discovery this year has been the use of those old beta-movie boxes that are now outdated. Video stores no longer use them and are giving them away. I use them for the storage of reading and math games. They are easy to label, and I can file them like books. The children have easy access to them in our classroom beta-library.

Maribeth Berliner, *Grade 3, Vermont*

I find the best materials you can get are individualized — that is, teacher-made. Kids love it! You can also make them to suit the child's abilities. It also lets you be creative. I have gone to companies to ask for donations. Parents also can become involved by asking their employers for donations. They do want to become involved. We also have places available to get low-cost materials like art supplies. Be creative and have fun.

Debra L. Mann, *Grades K – 5, Special Education, Massachusetts*

Enlist the help of parents, relatives, retired senior citizens, etc. Ask them to collect and save common items found in the home that might be recycled into classroom materials. Bundles of popsicle sticks make great place-value manipulatives. Magazines, especially ones with colorful photographs, can be put to use in classes for language arts, mathematics, creative arts, and more. A volunteer parent group may even come up with ideas you hadn't thought of!

Julie Abell-Victory, *Grades K – 12, Pennsylvania*

I carefully plan ahead for virtually every class project whether it be an art project, cooking, science, etc. When I have a good idea of my needs, I formulate a wish list during my open house for parents in the fall. I update the list frequently throughout the year. Parents have willingly kept me supplied with a wide array of materials for art – buttons, fabric, glue sticks, unusual odds and ends.

In science we have built solar ovens with donated materials. On some occasions, I have the principal print my wish list on the morning report. Other teachers and students in other classrooms often come through with some treasure. The beauty of this system is that when I get my $150.00 to order classroom supplies, I can use the money for real neat materials – unusual paints for art, new science materials, or hands-on materials. If it's going to benefit a child, I've found most people to be very generous and helpful.

Mary Ann Gildroy, *Kindergarten, Montana*

Have a class spell-athon, read-athon, or math-athon with parents, grandparents and friends sponsoring students for academic work. Money collected can be kept on account for supplies and field trip expenses. Offer extra credit for students who bring in materials on a scrounge list for a particular unit. Check *creative* fundraising ideas that don't ask families to buy things they don't need. Encourage business/education partnerships to provide specialized materials.

Cay Spitzer, *Grade 6, Colorado*

Use volunteers to help make classroom materials. Attend workshops and visit other classrooms to get ideas for materials you can make yourself. Look through catalogs of school supplies and figure out how you can make your own materials. Ask parents to help supply materials for your classroom. Some examples of objects parents have provided for me are: buttons, magazines, lids, and materials for art projects. Once you ask parents for their help in providing materials, they will come up with some things you can use.

*Marcia Stirler, Grades K – 3, Iowa*

---

Only a few of my twenty years of teaching found me lacking supplies and materials. I found that teachers need to use the resources of the community for help. Lumberyards where wood scraps to make geoboards, chart racks, etc. can be obtained for free; carpet and tile remnants for math and other projects; card and stationery shops have donated colorful bulletin board items. When paper is in short supply, overhead projector acetate sheets are great to teach from. As far as not enough rulers, scissors, etc., keep one set for every pod of four students to share. When it becomes the property and responsibility of that four to manage, they do a good job.

*Karen Morgan, Grade 6, Idaho*

---

Send a letter home at the beginning of the year, or even send it during the summer, asking parents to help provide certain materials and supplies. This is especially helpful in the science and art areas. Be sure to send a thank-you in a newsletter as the supplies appear at school.

*Faye Karna, Kindergarten, North Dakota*

---

The problem of insufficient classroom materials and supplies can be used to promote creativity and help us discover ways to use the things we have. If you need books, use the writing process to have students make their own books. Do science experiments with common household materials. Make use of the public library. Also there are several books on the market about free and inexpensive materials. Solicit help from parents and businesses. Be sure to take advantage of public television.

*Penny Parnell, Grade 3, Alabama*

---

I contact the appropriate administration. I borrow from other teachers, purchase a few things of my own, and seek out grant monies to help purchase materials.

*Velma Bell, Grades 3 – 5, Learning Disabled, Georgia*

---

This has never been a real problem for me. I try to work with what I have. If I don't have an item, I look for some sort of alternative. I collect and save everything from egg cartons to toilet paper rolls and shoe boxes!

*Sharon Kelly, Grade 3, New Jersey*

If grants are denied and the PTA is in the poorhouse, raise your own money through a classroom fundraiser. Inform parents of your problem and solution, then have the students select the products (classroom-made or commercial items).

Have a spell-athon or math-athon. Children will seek sponsors (they pay a certain amount of money for each correct item, or a flat fee), prepare the spelling test, and then collect donations after the teacher corrects the test.

Janet Mar Dong, *Grade 5, California*

---

As you come across things you would like to have in your class, write them down. When you have a list, hold a parent meeting and enlist help to make the materials. If it will make your job more successful and therefore more meaningful for you, spend the money and buy the materials for yourself.

Barbara Baker, *Grades K – 6, Hawaii*

---

Beg, borrow, and steal! Never throw anything away! Old game pieces, plastic shower curtains, egg cartons, save everything! Be creative. Parents love to donate; *a lot* of them cannot donate time, so they can save items (e.g., unused computer paper) while they are working.

Barbara Herman, *Grade 1, Oklahoma*

---

The key words to being an effective teacher are *be prepared* and *plan ahead.* Every week or two, I send a list home of "stuff" we are going to need for future projects. I also hang the list in the teacher's room. Most children love to bring "stuff" to school and we usually end up with what we need. I also plan activities using recycled materials that are always available – rocks, bread tabs, milk jugs, etc. These group activities spur more language development and understanding than silent activities.

Marilyn K. Emmons, *Grade 1, Montana*

---

Lists of materials and supplies can be sent home at the beginning of the year and parents can be asked to donate an item or items listed (if three or four parents each donate a box of crayons, there will be enough for a class crayon bucket, which can be shared by everyone). Many stores and businesses will donate old displays and surplus materials if asked. Many common household items can be utilized in the classroom. Margarine tubs and plastic food containers can be used for storing crayons, number and word builders, math manipulatives, and as paint tins. White construction paper can be completely painted by the students and the resulting colored paper used for art projects. Art paper can be cut in half to allow for more projects with limited supplies. Many math manipulatives, such as base ten blocks, can be made by the teacher. Magazines such as *The Arithmetic Teacher* contain suggestions for teacher-made materials. Many children's book clubs offer "points" on the basis of books which have been bought. These points can be used to acquire free books or teaching materials. Be creative!

*Anonymous, Elementary, Rhode Island*

---

Newspapers are a great supplement to your curriculum at minimum expense. Give each student a column of two or three paragraphs. Have them circle all adjectives red, verbs blue, underline compound words, etc. Underlining three statements and rewriting in question form is another activity. The advertisements are great for math activities. Cartoons are excellent for sequencing activities. Other activities include circling ten words in a story, then writing them in alphabetical order. The list is endless. Large newspapers are glad to help you supplement your curriculum with a list of activities such as these.

Recycling worksheets is another excellent practice. Laminate the sheets, then answers may be erased for the next student's use. After using reading comprehension sheets, cut the questions off the bottom, glue the story on construction paper and illustrate the story. Another activity is for one student to read the story, then write three questions about the story and give to another student to answer.

*Charlotte L. Hargett, Grade 1, Mississippi*

---

The library can be a good friend to the classroom. Check out as many books as you can to make the classroom literature-rich. The copy shops often have boxes of scraps. These scraps are great for writing stories. Investigate the types of grants that are available and apply for them. Garage sales can be a rewarding adventure. It usually boils down to your hand into your pocketbook.

*Mary Geren-Saggau, Grade 1, Nebraska*

## CHAPTER HIGHLIGHTS

Parents are typically most responsive to classroom needs. Use and reuse materials whenever possible and never throw anything away. Don't be afraid to enlist the help of others to acquire necessary items.

## CHAPTER 6

# Dealing with Individual Differences
## of Students

EVERY student is different. It's a fact you must recognize if you hope to promote academic achievement in the classroom. There is absolutely no getting around it, your students are going to present individual differences when it comes to behavior, commitment, motivation, learning levels, capability, and other qualities.

Given these conditions, and given that your ultimate goal is to provide success for every learner, you have quite a challenge. Success for each learner can be had, but only if you focus on one key word. That word is *individual.*

First, you must get to know, really know, each student as an individual. Secondly, each individual must remain as an individual. That means you don't make comparisons. Not to other students in the class or other classes, not to brothers or sisters, and certainly not to what's been designated the norm. Avoiding comparisons is not easy, but it is the thing to do. As you will find expressed by many of the veteran exemplary teachers, the application of learning styles may well be the method by which to best accomplish this most noble goal. Whatever method, strategy, or technique you settle on, time and energy invested on your part will be well worth it, and you won't be a genuine teacher until you have it comfortably tucked in your repertoire of instructional skills. Get started early on this one. You'll like the results.

---

Dealing with individual differences of students has become my personal specialty as a teacher. Seven years ago I began my own research in the area of learning styles. Nothing in education has become as valuable a tool to me as that knowledge. Now I assess each student's learning style in the fall, with parent-volunteer help. I have worked to create teaching strategies and multisensory techniques that will reach a wide array of individual styles in a large, diverse classroom.

Each spring I do an in-service for my parents on their child's style, so they can be more effective in dealing with homework and future classrooms that may not address their child's style. I have created a workshop called "The Colors of My Rainbow" to educate teachers and parents. I have presented this workshop over sixty times in Montana, Wyoming, and Idaho. Understanding learning styles made me a better teacher.

Mary Ann Gildroy, *Kindergarten, Montana*

---

One of the most important things to remember is don't compare students, siblings, or classrooms. Look for the good in each child and cultivate the stronger points. Students will need different levels of work, according to their abilities. It is important to conference with all students so they can see their progress, as well as their needs.

Pam Smith/Terry Carter, *Grade 6, Illinois*

---

In dealing with individual student differences, I always try to adjust the situation to fit the child. This year I have a handicapped student with muscular dystrophy, so he is unable to complete the same amount of work as the rest of my sixth graders. I cut assignments in half for him. I also do this with students with slower learning abilities. It serves no purpose to do forty math problems poorly when I can judge a student's work with only ten to fifteen completed well.

Ron Baysinger, *Grade 6, Indiana*

---

Throughout the ages, teachers have had difficulty dealing with the individual differences of students. Zero in on an activity that has real value to the students; one that some of the students can manage on their own so the teacher can focus on the majority. Clip articles of interest to the grade level from the newspaper, and when students finish the day's assignment, have them write letters to real people about real issues. Have a letter-writing center with numerous letter choices. Mimeograph the articles on a letter form. Letters to real people about real issues will be answered. This is a good motivational technique for students, and fun for all.

Shirley Spear, *Grades 4 – 6, Texas*

---

After teaching a lesson, some students work independently, while slower students are invited to work at a large table with me. That way it's easy to see problems and answer questions, or even reteach the smaller group if needed. I have also found that a student who has mastered a topic is an excellent helper for someone who needs extra practice. I always have enrichment activities, manipulatives, or computer programs available to challenge the student who finishes work early.

Shirley H. Goodemote, *Grade 1, New York*

---

Every child learns in a different way. For some children, learning comes easily while for others it is painstakingly difficult. Take each child where he/she is and go from there. Find the modality that each child learns best in. All children need to gain self-confidence by meeting success. We must have high expectations of all our students. We must be kind, patient, and nurturing with young children.

Sharon Conley, *Grade 2, Maine*

---

A teacher must take into consideration that each child is a different individual. This applies to everything from work, to attention, to discipline. Dealing with various discipline problems, I have to use several different approaches throughout the year. For me, to stick to one approach all the time gets no results from some of the students. I use a reward system and time out alternately.

Stacy Burns, *Grade 1, Mississippi*

---

Talk to your students as individuals and as a class. Make sure they are aware of their strengths—everyone has them—and weaknesses. Make sure they know that it is okay to have problems with some things, and that it is so with everyone, even adults. Issue the weak areas as a challenge, and offer lots of encouragement. Involve the entire class in working toward certain goals and have them help where they can. Be available to help after school if needed, and be in close contact with parents if there is a serious problem. If you are able to treat each student as an individual, and become familiar with them as people, not just students, it helps to know when and how to meet their individual differences.

Daria Ann Wood, *Kindergarten, Wyoming*

---

There is a wide range among students in each grade. Children come from different backgrounds and home environments which are dominant factors in their performance. Praise all work. Remember the A student isn't necessarily the student who worked the hardest. Touch on the visual, the audio, and the hands-on techniques when teaching. Have reinforcement and challenge work available to students in the form of learning centers. All students are gifted, but it may take digging a little deeper to pull it out of some students.

Eileen Bruton, *Grade 5, Wisconsin*

---

Respect all differences, culture, language, academic ability, home environment, etc. The teacher should allocate time for those students who have the greatest needs for that particular day. Allow students the opportunity to interact, thus learning from each other.

Kathryn Maloy, *Grade 2, Florida*

---

I believe that even in a classroom where three ability levels have been identified, there is a further breakdown of individuals whose needs have to be addressed. An instructional movement toward providing whole-group instruction has facilitated my efforts in meeting the specific needs of these individuals. Once I have completed my whole group instruction and have assigned practice work or a project to the class, I call together children whom I have identified as needing teacher assistance. These are the below-average students. The average students are asked to attempt the assignments on their own, but are instructed to ask for assistance as they find the need to clarify directions, identify unknown vocabulary, or repeat skills instruction. Above-average students almost never require assistance. They are able to complete enrichment projects and assignments in addition to the group work. I do this across all curriculum areas, but especially in reading. This method releases me for the extra time which is needed to do choral reading, ''big book'' reading, and reading from trade books with slower students. The more these students read, the greater their progress in developing reading skills. I also modify class assignments for students who take too long to process information, have poor fine-motor coordination, or have attention deficits. For these students, assignments are made shorter, some are eliminated, or directions are changed so that answers can be given in methods other than lengthy written responses.

Pauline McCollum, *Grade 2, Florida*

---

Get to know your students well. Each child has his/her own unique learning style. Observe each child and record how he/she prefers to learn. Parents can also give valuable insights as to how students learn. Focus on the strengths in each student, and introduce variety in learning. Your challenge is to create lessons which will take all learning styles into account. This may be accomplished by using manipulatives, listening centers, tactile learning centers, visual stimulation, and other methods of learning.

Becky Ann Erickson-Danielson, *Grade 1, Minnesota*

We should not treat a child like he or she is different because we do not want him or her to feel different. If we treat a child as different, the child may become frustrated and give up. We, as educators, are to find ways that will help overcome the individual differences, but keep individuals related to the whole class so no child feels put on the spot. The children can then look at each other as being equal.

Joe F. Coleman, *Grade 4, Ohio*

Talk about differences with students repeatedly. Discuss strengths and weaknesses. Give the message that we all have something to offer. Encourage peer tutoring whenever it is appropriate. Social skills as well as academic skills can be taught in this way.

Christine Shepard, *Grade 2, Utah*

First, appreciate the fact that each child is different. You must be convinced that every child can learn. Then, you decide how best to teach each child in your classroom. Find out what each child's ability is, and start teaching him on that level. Do not give too much too fast. Help each child to achieve at his own rate. Provide work at different levels of difficulty. Use different learning styles. Know your children and find out how they learn best. Then use that method to teach them.

M. Faye McDonald, *Kindergarten, Mississippi*

For students who are weak in a particular area, send home a copy of a former text no longer in use in your classroom. Assign short extra credit assignments the student can work on at home to improve his/her skills. Reward the student for his/her extra effort; maybe even allow the student to mark his/her own paper. Most students feel important correcting papers. Also the student will learn by seeing his/her mistakes.

Wendy Romano, *Grade 3, New Jersey*

First always consider that every child is different and special. When planning lessons, allow for the children who learn by seeing, and provide visual information. Encourage the children who learn best through hearing to verbally share what they've learned through the lesson material. Above all, plan for those who learn best by doing. Active planning will ensure that you meet the needs of the children in your class.

Sharon Kelly, *Grade 3, New Jersey*

We have a "Proud Person" of the week, where each student has a full week to be highlighted. The week is full of student-directed interviews, pictures (family backgrounds), favorite books, toys and even a favorite snack. To say the least, we learn an awful lot about each other. On their final day I have each student come and whisper something special about that person to me, and I write it down just as the child says it. We then read it and the list of "sweet thoughts" are taken home and shared. Parents have told me that they will keep it in their child's baby book for them to read when they are older, or when they forget just how special they are.

Barbara Herman, *Grade 1, Oklahoma*

Be prepared to provide work on every child's level. Give assignments that are cut in half for some and add a "what-if" problem on the board for children who finish early. Learning centers help, too. They give children an opportunity to practice and learn skills we are working on in our classroom. Another option is group tutors. Children who need a challenge can work with a group of slower students. Everyone benefits, the teacher has time to help someone else, accelerated students get to be in the spotlight, and slower children get the benefit of hearing a different explanation.

Marilyn K. Emmons, *Grade 1, Montana*

Observe your students and keep a record of certain behaviors which will help you to gain a personal understanding of their strengths, feelings of self-esteem, varying learning styles, need for structure, etc. In preparing lessons, plan for diversity so you will be able to meet the needs of all types of students. Try to include visual, auditory, and kinesthetic approaches. Plan for both individual and group work. Also, offer your students a choice of activities and projects. Finally, encourage peer tutoring, a situation which benefits everyone involved.

*Anonymous, Elementary, Rhode Island*

## CHAPTER HIGHLIGHTS

It is important to see each student as an individual. After all, each one is unique. You will find commonalities with which to group your students. Teaching should be individual as well as group oriented.

## CHAPTER 7

# Heavy Teaching Load with Insufficient Preparation Time

ALMOST all teachers, especially beginning teachers, find themselves with too much to do and too little time to do it. When this happens to you, don't be too hard on yourself. The inherent assumption in teaching is that you will *never* be fully prepared, because there is always room for improvement.

As with the other problems we address in this book, there are a few insights and strategies to help ease the pain. Things like scrutinizing your daily goings-on. You will find some things in your schedule that really do not need doing. Get rid of them. You will also find things other people — students or parents — can do. Shift those items to them. You will also find that a network of other teachers close to you, in the same building, at the same grade level, or in the same content area, can be an effective resource to shorten lesson preparation time.

Above all, be organized. If organization doesn't come naturally to you, develop the habit. What follows are suggestions you may find helpful.

---

I use cross planning — combining subjects when possible. I save all old plans. To me, organization is the key. Never touch a piece of paper more than once.

*Anonymous, Grade 6, Alabama*

---

Heavy teaching loads can only be changed through scheduling changes. If you feel you are running on overload, let your colleagues know, then talk with the administration. Perhaps rescheduling will help.

Historically, teachers are overloaded with preparations, corrections, reports, committee work, planning, testing, curriculum evaluations, etc. Keep yourself organized and focused. Be sure to leave time for yourself and of course your family. Quite frankly, that is teaching — overworked and underpaid, and rewarded from within.

Dorothy Hartson, *Grade 6, New Hampshire*

If you are fortunate enough to be in a large school where there are several classes of the same grade, you have some options. You can team teach, with one teacher preparing and teaching social studies and the other science, etc. You can each develop a unit and then share your plans. Just sharing a planned worksheet or an evaluation will be helpful. You can also relate to other grade level personnel in neighboring schools and arrange a swap of materials and ideas. Use parent volunteers to go to the instructional resource center to mimeograph/xerox, and laminate visuals. Just remember — you can't do everything at once. Just strive to do the best you can under the given circumstances.

Betty Robeson, *Grade 3, Maryland*

As devoted teachers, we all know that the needs of the students, both academically and psychologically, should be met first. Even so, you need to go easy on yourself as well. If you are extremely overburdened and rushed, you will more likely overlook the needs of some students, rather than meet them.

Lesa Carroll, *Grade 5, Missouri*

Too many kids — too little time to *organize!* Eliminate the need of students asking you when they can decide for themselves. Establish when pencils can be sharpened, bathroom trips made, and what materials may be used freely as they need them. Allow students to get help from other students — peer tutoring. Answer questions only when a peer has been asked first and cannot help. Do wall charting so students have visual reminders of things they need to work on. Establish self check areas so students get instant feedback and do not have to wait on the teacher. Help students to become more independent, not more dependent, on a teacher.

Barb Parmenter, *Grade 4, Michigan*

My solution to this problem will probably not be very popular, or even an option for most teachers, but it works for me. I arrive at school between 6 and 6:30 A.M. and remain there until 4:30 or 5 P.M. I still take work home. We have no in-school preparation time, so this is the only way I can keep my head above water.

*Anonymous, Grade 4, Utah*

With the teaching load, conferences, faculty meetings, and reports, preparation time is very often limited. Extra hours of work after school are a reality to any good teacher. However, there is one way to help ease the pressure of insufficient time. Every teacher should have a "buddy teacher" whenever possible. The simple sharing of ideas and thoughts can improve the quality of your plans, and speed up the planning process. When teachers of the same grade level work as a team, the result is a large pool of ideas, consistency in the curriculum across the grade level, and highly productive use of time.

Kathryn A. Smith, *Grade 2, South Carolina*

---

Make a list of your main priorities for the year. Be honest and cutthroat about it. See what is in your daily and weekly schedule that is sabotaging these priorities. Eliminate frills and nonpriorities. If it is very important to you that the students thoroughly understand their math concepts, then spend your time preparing meaningful activities with manipulatives, do not use your time to run off drill sheets. Do your top five priorities constantly, rotate in your next five, and save your others for another year when you have a lighter load or more time.

Leslie More, *Grade 2, Washington*

---

Since the skills are taught yearly, I find that keeping very good files of the lessons taught, and the activities related to each lesson, helps. I can always refer to these files during the following school term. Of course, revisions should be made to keep files updated.

Linda N. Stevens, *Grade 1, Mississippi*

---

This is a daily problem. Document what you do and set up units. Each year your job gets easier. Updating takes less time. Learn to go without sleep!

William Andrews, *Elementary, Massachusetts*

---

This is a very real problem for most teachers. More and more demands are being placed on teachers, and more is being wedged into the curriculum. I have sometimes felt very isolated in my efforts, as though there was no way I could complete all that was assigned to me. The solution was to do team planning and teaching. At least once a month, the third-grade team of teachers gets together to plan units and consolidate resources and ideas. It is refreshing to have so much collegial support. Not only do the teachers benefit from such a strategy, but so do the children. Things that may not have been possible before, are now possible because of combined efforts.

I also make use of classroom mothers/fathers/guardians to help with some of those time-consuming tasks. They love to be a part of the classroom.

Maribeth Berliner, *Grade 3, Vermont*

---

Nothing is worse than no prep time. These last three years in Boise are the *very first time* in twenty years I have had it. Extremely good organization is the key. Planning ahead in unit blocks helps. Enlisting the help of parents for running off papers, cutting out things, etc. is vital. I keep a folder that contains the week' s work. I make sure I have all materials on-hand one week before the lesson.

Karen Morgan, *Grade 6, Idaho*

---

Our school has developed a high school incentive student program. The students who have too much study hall time can earn half a year' s credit by coming into our classrooms to wash boards and desks, ditto, staple papers, correct papers, help with parties, read orally to the children, help with art, put up bulletin boards, etc. The students come into our rooms while we are teaching and do the duties we assign them each day. The students must be responsible for being there on assigned days and hours, or they must let the teacher know where they are. Some teachers have the students sign a sheet each day and state what they accomplished, and give their comments on the assigned tasks.

Faye Karna, *Kindergarten, North Dakota*

---

Use some time on the weekend! *We all do!*

Barbara Baker, *Grades K – 6, Hawaii*

---

Enlist parent and other volunteer help. As a first-grade teacher, I know most parents of young children are eager to be involved in classroom activities, and the children love having mom take part and help out. Volunteers can be a great asset by tutoring slow students, grading papers, working with small groups, reading to students, and taking the class to lunch, thereby allowing you, the teacher, more planning time.

Other ways include purchasing rubber stamps to save time writing messages over and over (e.g., ''Please sign and return,'' ''Please correct and return,'' etc.). Also, allow students to correct their own papers when you do not intend to record the grade. Before doing this, have them color all answers with a yellow crayon. By doing this they will be unable to change or erase answers.

Charlotte L. Hargett, *Grade 1, Mississippi*

Volunteer parents, high school students on work study, and interested community citizens who volunteer in the schools have become my lifesavers. Some years I juggle activities for as many as twenty volunteer helpers. (I teach two sessions of kindergarten.) Volunteers help me with everything from secretarial activities (copying materials, typing, stapling packets), to preparing materials for art, science, and social studies, to helping with my music program by singing with us or playing a musical instrument to accompany us, to individual remediation activities and creative writing, and even helping me clean and rearrange my room.

Learning how to work with so many individuals involves careful planning on my part. I need to have materials ready and often need to have samples prepared. Volunteers free my time to do more creative planning and teaching. They often become cherished friends to the students and me, and are wonderful public relations people in the community because they see and understand educational needs from a clearer perspective. They are great!

Mary Ann Gildroy, *Kindergarten, Montana*

---

Organization! I run copies of worksheets and tests needed at least ten weeks ahead and keep them on file. That way these things are available to use on a daily basis. Always save and file a copy of materials to be used for another year. I file them in order of use throughout the year. I have also found older students who have the time, and love to help in the room. They are excellent at preparing materials for art projects, small group reinforcement games, or a few minutes of individual attention.

Shirley H. Goodemote, *Grade 1, New York*

---

Making good use of planning time and being well-organized are essential ingredients. In our second grades, we team teach for literature and health. We share ideas, materials, and units. Teaching is a labor of love. Our reward is our students' success and happiness in learning.

Sharon Conley, *Grade 2, Maine*

---

Set your own limitations and be aware of them. Each person can only do so much and no more—just always do your best. Prior planning is the most important part of each lesson. Each year gets a little easier, and if you spend the time learning how to plan really good lessons, then you will be able to plan great lessons in relatively little time.

Sharon Kelly, *Grade 3, New Jersey*

## CHAPTER HIGHLIGHTS

It is vital to keep good records of your plans, and then make additions or deletions where needed. Don't be afraid to utilize parent or community volunteers to do routine tasks such as xeroxing, readying materials, or grading papers. Nothing short of great organization is needed!

## CHAPTER 8

# Relations with Colleagues

WHEN you take your place in your school, you become a part of a system. The system includes a tangible student body, an intangible esprit de corps, an intangible culture, and a very tangible cadre of colleagues. All are important, and help give a school its identity. But you and your cadre of colleagues are responsible for, and are what determines, the operational success of the school on a day-to-day basis.

Your colleagues are your family. They are also your team members. You receive from your team and you give to it by sharing your knowledge, your energy, your enthusiasm, and your caring. When all this takes place, what you will have at your immediate avail is the best resource base you could ever hope to have. That in itself is a good enough prize, but what you will also discover, much to your delight, is that this resource base, this teamwork, this esprit de corps will pervade your classroom, and you and the system will be better off for it.

---

Keep these very important relationships open and friendly. Don't allow them to get competitive. Colleagues can save your sanity in your first years. Try to keep your relationships positive and professional, or in other words, don't allow them to become gripe sessions. If you socialize with good teachers, you'll get many good ideas, and be able to have a sounding area to try out your ideas.

Sharon Kelly, *Grade 3, New Jersey*

---

Always try to do whatever you can to help the people around you. Be positive and look for the good side of things. Be open-minded and relish a sense of humor. If you are cheerful and cooperative, others will respond in kind. Find out about your colleagues as *people,* not just as teachers. They will be long-term associates who affect you in many ways. Good relationships are worth the effort.

Shirley H. Goodemote, *Grade 1, New York*

Refuse to be a part of gossip about other colleagues and their classrooms.

Barbara Baker, *Grades K – 6, Hawaii*

It is tricky to talk about relations with colleagues because each situation is different. However, what I feel is the most important thing to remember is that we are all in this business together. Support and flexibility are the keys to successful staff relations. Being able to share ideas and give constructive comments to each other is also very important. At the beginning of the year, we have staff development workshops where there are opportunities for staff to get better acquainted with one another. For example, this past fall, we had a ropes course where we worked on team building, self-esteem, and trust. It was great fun! It is important to note that many of these experiences are made possible by a supportive administration which is sensitive to the needs of the faculty. We have such an administrator.

Maribeth Berliner, *Grade 3, Vermont*

Realizing that we have to deal with many different personalities, we should always be approachable. We have found that if everyone attempts to work together, it will create an environment that is always pleasant. I find that always treating others as I would like to be treated has proven to be the best policy. I also share materials and try to be helpful to anyone I can. We have to keep in mind that our main business is educating children. We should conduct ourselves as professionals at all times.

Linda N. Stevens, *Grade 1, Mississippi*

Building a positive work relationship with colleagues requires flexibility. With teaching comes those unexpected schedule changes, class cancellations, and conflicts in availability of materials. Teachers learn quickly to work out these usually unavoidable problems with colleagues by being accepting of change, and flexible enough to adjust quickly.

A willingness to share is also a key factor in developing relationships. Sharing ideas with others provides an open invitation for them to share as well. Sharing builds success. A good teacher is always in search of new ideas to make the classroom a more motivating and instructionally successful place for children.

Kathryn A. Smith, *Grade 2, South Carolina*

Teaching is never an easy job, but it can be made easier and more pleasant if you develop a good relationship with your colleagues. They can be your most valuable resource in solving problems, working through new ideas and giving you support. Be willing to share your ideas, materials, and time. A good idea dies unless it is shared and used, so be generous with yours. Conference with coworkers often. Plan together and work together. Share responsibilities and assignments.

*Anonymous, Grade 4, Utah*

---

Toward the end of the school year, have all the next-year teachers (of your students) schedule a few minutes to come and read their favorite children's book to their upcoming students! It is a nice way to introduce the students' possible teacher. It is less intimidating to students, because their new teachers are coming on the students' familiar territory. Peer teacher relationships are also strengthened through this type of activity.

Barbara Herman, *Grade 1, Oklahoma*

---

Don't feel like you are in competition with other teachers in your grade. We are all trying to do the same thing – educate our children the best way that we can. Our kindergarten teachers meet frequently and discuss units and objectives which we all teach at the same time. We periodically meet with our principal to discuss any problems or needs that we might have. We all share ideas, projects and materials.

Faye M. McDonald, *Kindergarten, Mississippi*

---

For a beginning teacher, there is usually someone to rely on when there are questions. It doesn't have to be someone that is an old hat at teaching, but someone that you feel comfortable with asking anything about school. Remember that you are only a small part of the big scene, and be open and flexible to others' ideas. There is no one tried-and-true way to deal with teaching. It takes trial and error. Practice, and don't be afraid to make a mistake. I always want my classes to know that I am human, too.

Stacy Burns, *Grade 1, Mississippi*

---

Good relations with colleagues make for a good learning environment. Have trouble getting students to use their subject-area vocabulary outside class? Meet with the writing teacher and suggest that the students use their subject-area vocabulary in their writing class. The words can be highlighted with a yellow crayon, and the subject-area teachers can devise a reward system for the students who participate. It's contagious!

Shirley Spear, *Grades 4 – 6, Texas*

A sharing and caring faculty is worth a million. Openness and honesty need to be established in every faculty. An early-morning smile and laughter serve as good forms of medicine. A bulletin board for the sharing of ideas and materials is very helpful. Listening is a key to knowing your colleagues – sometimes it is better than giving suggestions. Don't pass judgment.

Pam Smith/Terry Carter, *Grade 6, Illinois*

When you come to the realization that you could use more help or need ideas, do not be too proud or too embarrassed to go to a colleague and seek help. You'll be surprised how receptive others are to your plight – they were in your position once upon a time.

Janet Mar Dong, *Grade 5, California*

If your building or district does not have a mentorship program, find a colleague who is willing to serve as your professional tutor. Meet regularly, keep notes about your questions, and record suggestions. Invite administrators to observe you in action. Arrange to visit other classrooms during your planning period to see how others handle discipline, grading, etc. Make contact with teachers at the same grade level. Watch, listen, and learn from everyone!

Cay Spitzer, *Grade 6, Colorado*

A high degree of communication and collaboration among colleagues can keep relations positive. Small problems or irritations can be handled before they become disproportionate. You don't have to be best friends with colleagues; you do need to have respect for each and accept his/her individuality. Also, teachers go through states of professional growth that might be classified as the *novice* teacher, the *developing* teacher, and the *expert* teacher. Each stage has its own strengths and weaknesses. One colleague can help another to capitalize on the strengths.

Julie Abell-Victory, *Grades K – 12, Pennsylvania*

Colleagues are very important in your career. They are a source of experience and wisdom. I deal with my colleagues on a one-to-one basis, because of the various types of teaching styles, attitudes, and ages. Personally, I listen, and I keep whatever is said confidential, unless the person wants the information to be said. I also give my opinion when needed. However, I state it as my opinion. You have to work as a team, and become a team player. Opinions are expressed in a positive or negative statement. It's important to respect them as much as they respect me. I also try to greet everyone whenever it is possible. Remember, every other teacher is a valuable person who can help you out in many ways.

Debra L. Mann, *Grades K – 5, Special Education, Massachusetts*

---

The teachers in my building who are the most popular with peers seem to adhere to the following.

- Make other people feel good about themselves.
- Try to see things from the other person's point of view.
- Do your share of the work.
- Be willing to help without expecting recognition.
- Be dependable.
- Use your ears twice as much as your mouth.

Barbara J. Bell, *Grade 4, Pennsylvania*

---

This can be one of the most challenging of teachers' problems. The best starting point is to wear a friendly smile and have a good sense of humor. One needs to be able to laugh at oneself. Offer a compliment on a nicely done bulletin board, room decoration, or article of clothing. When discussing a problem with a student, offer a suggestion in a kind and caring way which doesn't come off as though you know it all. Share materials, equipment, and ideas rather than hoarding things so that no one else has access to them. Be ready to share ideas that you have acquired from lectures, workshops, or classes you have attended. Be a friend and you will have friends. Be a professional colleague and you will have many professional colleagues to work with, and your experiences in education will be wonderful. The classroom can be a very lonely place if you try to go it alone. Take some time to visit with your colleagues in the teachers' lounge over a cup of coffee or can of pop. You will be surprised what you can learn in a very nonthreatening way.

Frankie C. Bly, *Grade 5, Minnesota*

## CHAPTER HIGHLIGHTS

Be receptive to your colleagues' ideas and listen with enthusiasm. Good plans should be shared. Always be positive in dealing with your fellow teachers.

## CHAPTER 9

# Dealing with Problems of Individual Students

NINETY-NINE out of every one hundred teachers will have problems with individual students. There are three or four very basic strategies to help you deal with individuals. First, focus on *individual* needs. Most of us have an inherent tendency to focus on the class.

Next, develop and maintain a good set of records on your students. These files don't have to exude sophistication — all you need are notations of what works for each of your students, and what each responds to in various instructional settings.

Also, stay current in the field of learning strategies. This is one way, maybe the best way, to show your commitment to, and endorsement of, dealing with problems of individual students.

A final basic strategy is the development of a personal data bank of ideas, techniques, and options to give you the firepower you need to take on each and every one of the classroom problems of your individual students. Here are ideas which have worked for other teachers.

---

The important point to remember is the word *individual*. Every child is different, and should be treated accordingly. What works with one child might not work with another. Children these days come to school with a lot of excess baggage. We have no control over those outside influences, but we can help children in how they respond to their problems. You need to show the child that you are interested in him as a person, and that you care about what happens to him. Enlist the help of parents when you can. With parents, teacher, and child working together, chances of success are a lot better. If the problem is really serious, don't try to take it on by yourself. Compile a list of outside resources so that you can readily consult it when the situation arises.

Nancy Mahloch, *Grade 6, Nebraska*

---

Use the time in the morning before school starts to work individually with students needing extra help. It also helps to match a strong student with a student at risk when reading in content areas such as social studies and science. Another technique that seems productive is to have the students in a slower group help younger students learn new skills. This provides the older student helper with review and also builds his self-esteem.

Barbara J. Bell, *Grade 4, Pennsylvania*

---

The developmental stage of each student needs to be taken into account in order to provide meaningful educational experiences for all. A middle school child who is still in the concrete stage of intellectual development (Piaget) will have some difficulty with abstractions in mathematics. An age cohort in the same class may have made the transition to the formal stage of intellectual development, and may be able to view a task from a much different vantage point. Differences in stages of intellectual development may be particularly obvious with students who are six to eight years old, and again, with those who are ten to twelve years of age.

Julie Abell-Victory, *Grades K – 12, Pennsylvania*

---

Develop an assistance team policy with the other teachers. Its function is a joint effort of the teachers to develop a concept of the student problems, brainstorm solutions, plan interventions, and conduct follow-up meetings to evaluate student progress. Develop some goals and a definite procedure. Into this policy is also listed the personnel such as team members, support personnel and parents, and how each is involved. Once a student is referred to this team, a file is started for him and signatures are secured for the team members, coordinator, and school board president. The teacher who asks for this assistance gets good ideas and support.

Faye Karna, *Kindergarten, North Dakota*

---

Very effective for me over my twenty years is an individual contract for each child who is having trouble functioning in the classroom setting. It takes care of behavior problems as well as problems in the academic area, or with study habits. No matter what the problem, I find the time to meet with parents, assess the problem with them, conference with the child, etc.

Karen Morgan, *Grade 6, Idaho*

---

Students will respond to you if you have specific actions that they can do to change the situation. Caring is important. Always be honest.

William Andrews, *Elementary, Massachusetts*

For students who have difficulty organizing their time and behavior, I use a weekly chart taped on their desk. Divide the chart into four time periods per day. Use smiley faces to indicate when the student works well, and frowns when behavior is inappropriate. Have students set a reward for fifteen smiles—can accumulate from week to week. Ten frowns and the student is sent home—frowns *do not* accumulate from week to week. [See chart below.]

Barb Parmenter, *Grade 4, Michigan*

Student's name

## STUDENT CLASSROOM BEHAVIOR REPORT

|         |            |             |            |           |
|---------|------------|-------------|------------|-----------|
| MON.    |            |             |            |           |
| TUES.   |            |             |            |           |
| WED.    |            |             |            |           |
| THURS.  |            |             |            |           |
| FRI.    |            |             |            |           |
|         | 8:30-9:45  | 10:00-11:05 | 11:55-1:00 | 1:15-2:45 |

☺ means student's behavior was satisfactory to good.

☹ means student's behavior was unsatisfactory.

Weekly copy of behavior report sent to parents, principal, school counselor, and teacher. This report does not mean that all work was completed.

Remember that every child will have individual needs. They only become problems when needs are not met. Try to keep a daily classroom log in which you record classroom situations and events. Keep a folder for each student that identifies special needs, and your plans to meet those needs. You may also find that you need to keep a separate daily log for a particular student.

Lesa Carroll, *Grade 5, Missouri*

Many children arrive from a home situation that was unpleasant and/or stressful. Learn to know each child by reading the cumulative folders, talking to their previous teachers, meeting their parents, and listening to their stories. Accept each child as an individual and remember that the parent(s) sent you the best they have. If a student becomes disruptive, send him/her to a designated area for ''time out'' (a cooling down period). This may be a place in your room or in a neighboring classroom. If the interrupting persists, try to ignore the unacceptable behavior and give positive comments and your attention to those who are being good classroom citizens.

Betty Robeson, *Grade 3, Maryland*

Loving the individual students—that's teaching. I always use all the experience I have, all my ideas, and if that's not enough, I seek the help and advice of the school's experts. A school has many people—nurse, guidance counselors, principal, assistant principal, special education staff, librarian, specifically teachers, colleagues at other levels—who can help you solve a problem. Use them. Depending on the situation I select my advisor appropriately. Also, to help with individuals, I keep current with new teaching strategies and materials. I attend workshops, conferences and institutes to improve myself. Keeping yourself up-to-date as an individual helps you instruct the individual.

Dorothy Hartson, *Grade 6, New Hampshire*

For students who have trouble with organization, completing work, or even behaving, keep a weekly behavior modification chart tracking the child's daily success. Briefly record the child's daily progress under each heading and send the chart home at the end of the week. Parents can reward the child for earned success. Parents will also know the child's weaknesses and can set goals for improvement.

Wendy Romano, *Grade 3, New Jersey*

In today's society, students seem to be burdened with more responsibilities than seventeen years ago when I began. Because of this, students are facing more problems at an earlier age. I put aside time each week to talk with each student about any problems that may arise, be it in school, with family, or with their peers. If there are no problems, we just "chew the fat," so they become more comfortable talking to me if a need should arise.

Ron Baysinger, Grade 6, Indiana

I never read the files of my students until I have formed my own opinion of the student and his/her problems. This allows me to form my own opinion on every child, and to develop my own plan before I am influenced by the files. I encourage parents to meet with me to tell me what has worked in the past with the child and what hasn't worked. I try to involve the parents in the final plan for helping the student. It helps the child when she/he realizes everyone is concerned about the problem. I also help the class see that not everyone in the class will be treated the same way, because we all have special needs, and I want each child to work to the best of his/her ability.

Marcia Stirler, Grades K, 3, Iowa

## CHAPTER HIGHLIGHTS

Remember what works for one student may not work with another. Each child comes to class with a different attitude and set of values. Be willing to listen and make the student comfortable talking to you. This may help when problems arise.

## CHAPTER 10

# Assessing Student Work

THE primary relationship of student work assessment to classroom teaching is evident – we assess student work to gauge the students' mastery of the material. What may not be so evident is that we can use student assessments to gauge the effectiveness of the instructional program. When you realize that a child is not learning, you need to figure out what to do. Here are some thoughts:

- Assess the whole child. Don't just assess long-term memory, or immediate recall, or comprehension, or application – do it all. Each of us has a performance strength which needs to be identified.
- Assessment should be an ongoing, daily process. We all have good times and bad. Be careful not to skew assessment and misrepresent the learner. The Friday test, for example, is an isolated assessment and chances are the results are not truly representative of every learner's overall performance potential.
- Vary your assessments by setting (classroom, lab, field trip), discipline (science, math, language arts), involvement (self-study, partners, small groups), and instrument (criterion-referenced, norm-referenced, commercially made, teacher-made).

Good assessment can help each student shine.

Student assessment should be comprehensive. It should have no boundaries. Give yourself credit for your own ideas and put them to use. Ultimately you will discard some of the techniques – yours as well as others – but you will develop a functional repertoire of assessment techniques. Here are some techniques that have worked for our exemplary teachers.

---

I use a variety of tools which include teacher-created tests and/or quizzes. The emphasis in my classroom is on daily work, and I take many scores. The tests are averaged in, just like a daily score, with no more weight or importance than a daily assignment. It's good to use commercially done tests from the textbook company, but feel free to modify them to fit your teaching style. I use oral quizzes/tests. In getting ready to assess students at the end of a unit, I outline the test points on the whiteboard. I require them to copy the notes from the board onto their notecards. This serves as a review device which can be taken home, plus I use the activity to teach outlining skills.

Frankie C. Bly, *Grade 5, Minnesota*

---

Plan to have students grade some papers in class for immediate feedback. For behavioral and process objectives, record the objective, then mark the roster as you observe each student's daily activities. Writing and art are particularly good for developing portfolios to show a year's progress. In art, provide students with the criteria and have them judge their own work. This gives two grades for each project. Peer critiques may be helpful for older students.

Cay Spitzer, *Grade 6, Colorado*

---

Assessing student work, and, in particular, the writing assignment, can be done objectively by setting specific guidelines that all students can master. Tell students they will be assessed on the whole writing experience as opposed to bits and pieces such as spelling, capitalization, and usage. This frees the student to write creatively as opposed to focusing on error-lessness. In assessing the writing assignment always leave the option for all students to improve their grade. Make a deal! Make a deal with the students to take their papers home to revise and edit for a higher grade. Works every time!

Shirley Spear, *Grades 4 – 6, Texas*

---

I give students plenty of practice and chances to get it right. I take plenty of grades. I feel the more practice students have the better chance they'll all have to get the idea. Admittedly some children just don't catch on, but I know they have been given ample opportunity to improve.

Stacy Burns, *Grade 1, Mississippi*

---

Assessing students' work can be very time consuming if you let it be. I have learned that everything does not need to be corrected and graded by the teacher. Children can be taught to help by checking some practice pages themselves. Teachers can then do a quick recheck. Test scores usually give a good indication of a child's understanding of a concept. Don't underestimate teacher judgment. This is often more accurate than scores. Don't be afraid to use it.

*Anonymous, Grade 4, Utah*

---

Encourage students to assess their own work as part of the evaluation process. After they have assessed their projects, allow them time to increase the quality of the project if needed.

*Barbara Baker, Grades K – 6, Hawaii*

---

For each academic subject, a caption sheet is made that states the goal, the criteria that support the goal, and the opportunities by which the goal can be attained. This caption sheet accompanies the child's work that exemplifies growth, and is put into the child's portfolio. The portfolio is used in conferences with the child and the parents. This tool can allow the child to have a say regarding what goes into the portfolio.

*Mary Geren-Saggau, Grade 1, Nebraska*

---

The students are the ones that need to evaluate their work. Samples of the students' work need to be best papers, tape recordings of readings, stories they've written, etc. Have conferences with the students often, and always have a conference sheet that you fill out together with three headings: *what I've learned, what I need to learn next, when accomplished.* This helps kids be honest with themselves and see learning as a spiral, not a dead end road. It helps self-concepts because they don't need to hide areas of weakness.

*Leslie More, Grade 2, Washington*

---

Immediate feedback motivates students to work harder, and gives the teacher a measure of success in reaching daily objectives. A teacher must trust a student to grade homework and some daily class assignments, or the work load would be impossible. It is imperative to grade some students on ability coupled with effort. Continual failing grades for the student who tries can destroy a student's desire to learn.

*Helen M. Martin, Grade 4, Georgia*

---

Your most frequent assessing will be on daily assignments. This will give you a good idea of how students are doing on daily objectives. Chapter and unit tests will help assess how well students retain what they've learned. If more than thirty percent of your students do poorly, or fail specific objectives on a chapter or unit test, it would necessitate going back and reteaching those objectives.

*Anonymous, Grade 4, Michigan*

---

In this era of change in education, the trend is moving away from pencil/paper assessment to evaluation based on student participation and outcomes. The child, at whatever level of learning, needs to be able to demonstrate his learned knowledge. In addition, he needs to be able to demonstrate he knows how to apply that knowledge to a desired task.

Assessing students may mean moving from tests to products or projects. It may mean using a portfolio of compiled information – several forms of assessment – to place the student compared to a standard.

In addition, assessment will need some rubric of expectation. This needs to be clearly established and known to the student so they can strive at their level of performance.

*Sharron Baird, Grades 2 – 4, Oregon*

---

The National Council of Teachers of Mathematics has suggested guidelines which could be incorporated into any curricular area. Assessment should be integrated with instruction. A variety of types of assessment should be used; observations, student interviews, and portfolios are but a few examples. The assessment data that is gathered should be used to both help students learn, and to improve the curricular program. Assessment is much more than a paper/pencil test which is completed in forty-five minutes.

*Julie Abell-Victory, Grades K – 12, Pennsylvania*

## CHAPTER HIGHLIGHTS

Be sure the students and their parents know how you grade, and what your criteria are. Giving feedback lets your students know where they stand. Let your students help with the grading when possible.

# SECONDARY EDUCATION

Secondary education usually includes grades seven through twelve, although middle schools serving students in grades five through eight or even nine tend to blur the distinction between elementary and secondary education.

Learning behaviors change as students move from the elementary to the secondary level. These changes impact the learner, they impact the teacher, and they impact the priorities of entry-level teachers. For instance, the veteran elementary teachers contributing to this book indicated that their most pressing issue is relations with parents. The secondary level teachers suggest that their most pressing issue is classroom discipline. Elementary teachers ranked classroom discipline as their second concern. For secondary teachers, relations with parents only ranks third.

The point is that classroom issues, the inner workings of teaching and learning, change as students move along the academic continuum. Being sensitive to these changes should help the entry-level teacher have a successful first year in the classroom.

Part of being able to achieve success in any venture is knowing what's ahead. Here is the order in which our secondary teachers ranked their most important issues:

- classroom discipline
- motivating students
- relations with parents
- classroom management
- assessing student work
- relations with colleagues
- dealing with individual differences
- problems of individual students
- heavy teaching load with insufficient preparation time
- insufficient materials and supplies

As you read through the chapters that follow, you will be provided with many strategies, ideas, procedures, and techniques that other educators have found to be successful and well worth passing on. In almost all instances you have the unique opportunity to initiate a line of communication with the contributors should you wish to pursue the ideas presented here.

## CHAPTER 11

# Classroom Discipline

ALL classroom teachers have to deal with the problem of classroom discipline, but none have to take it more seriously than those who teach at the secondary level. Here is where the tenets of classroom discipline are given their toughest tests. Secondary teachers probably encounter precisely the same behaviors their elementary counterparts do, but the behaviors are generally more magnified, more intense, maybe even more polished. Accordingly, the responses to these behaviors, and the measures to prevent them, must be more polished than those at the elementary level. Secondary educators as a group share a number of classroom discipline basics. Some of them are as follows.

- Put your rules in operation on day one—but don't overdo it. You can be guilty of overkill and you'll be the ultimate victim.
- Take a cue from the Boy Scout motto—"Be prepared." If you are prepared for your classes and your students are involved (not just being kept busy), you'll thwart most disruptions. Good teaching begets good discipline.
- Have a good disposition. Show your students that you like your job. Try to be at or near the door when your students come into the room, and provide a variety of greetings. When your students leave, say good-bye.
- Be firm. Show that you're in control. But be sure to apply your discipline across the board. Never be guilty of having "pets" or favorites. Any given infraction of one of your rules warrants the same disciplinary response—no matter who commits the infraction. One note of caution—discipline only the offending student, not the whole class. Remember, it is no fun getting beat up for something you didn't do.
- The things to avoid are: actual confrontations with students; existing problems, like pairing up a couple of rowdy problem students; and "blowing up" in the classroom—the consequences last forever.

Now give serious consideration to what the veteran exemplary teachers offer as ways to promote classroom discipline.

---

I have the students arranged in a seating chart alternating boy-girl. I confer with the teacher they had the previous year to find student combinations that should be separated. Classroom rules are posted and discussed the first day of school. They are: be on time; be respectful; and do your work. I believe if the students sense that their teacher genuinely cares for them, wants them to succeed, and is interested in listening to their concerns, they will have respect for their teacher and behave accordingly. I smile a lot. I have a good lesson ready with no idle time, and planned work for them to do.

June Enowski, *Grades 9 – 12, Missouri*

---

The primary way I achieve discipline in the classroom is to constantly take the student into consideration. If you discipline in such a way as to remove the sense of conflict, things run smoothly. Respect their rights as people. Never discipline in front of others if it is at all possible, and be as consistent as possible.

Dan A. Fenner, *Grades 9 – 12, Pennsylvania*

---

Go to a disruptive student and speak quietly to him/her as the usual recourse. Avoid confrontations if possible. Utilize backup systems – department chairperson, parental contact, etc.

David Hungerford, *Grades 9 – 12, New Jersey*

---

Post five clear rules as needed and where everyone can see them. It is easiest if you have one consistent rule, e.g., do not disrupt our class. This gets rid of petty complaints and arguments from students. If they are disruptive they are wrong – no questions. Remember, the busy student does not have time to entertain thoughts of mischief. Use action activities and cooperative learning techniques to best utilize their energy and social tendencies.

Kimberly Sturgeon, *Grades 9 – 12, Virginia*

---

During the first two days of school, I review my expectations for the year, including discipline. Each student receives a copy of what is expected. This is to be reviewed by the students and their parents with a signature by both acknowledging their understanding of my rules. They are also tested in class on their understanding of the rules. If a problem occurs or reoccurs I first refer to their signed copy of the rules. Second, I involve the parents. Third, I involve the administration. I rarely have discipline problems because I move about the classroom as I teach. I rarely teach from just one specific place. I encourage a good working relationship with the students. They know I am the boss. They also know that I care deeply about each one of them.

Paul C. Worthen, *Grades 9 – 12, Arizona*

---

Establish a set of classroom rules and post them on the bulletin board. Inform the students of the rules. Discuss them thoroughly during the first days of school. Have a quiz on them. When a discipline problem occurs, handle it immediately. Separate the child from the group, if possible, for a one-on-one discussion. Return the child to the group and continue the lesson.

Brenda Clemons, *Grades 10 – 12, Virginia*

---

Be firm! From the beginning you must be in control to gain their respect and lead them the rest of the year. You can *always* loosen up, but it is almost impossible to become more strict.

Susan LaPeer, *Grades 9 – 12, Massachusetts*

---

You must *earn* respect. Always be punctual, organized, sharp, fair, dependable, and professional. Be respectful and pleasant, but remain authoritative. Present an agreement (positive expectations *and* promises) on the first day; enforce it consistently. Handle problems on a one-to-one basis. Keep students busy with meaningful tasks and work *with* them. Rarely raise your voice.

Elizabeth Bishop, *Grades 9 – 12, Michigan*

---

Prepare a short list of rules. Present the rules to the class on the first day of school, and discuss the importance of having rules in a civilized society. Be consistent about enforcing the rules. Detain rule breakers after school and counsel with them. Avoid taking disciplinary actions until you have discussed the matter with the student. Disciplinary action without counseling is a waste of time and effort. Always base your decisions on facts and logic rather than emotions.

Abbas Saberali, *Grades 7 – 9, Texas*

Be on time; be ready; and have enough work organized so everyone is challenged for the entire hour.

Sally Tanner, *Grades 10 – 12, North Dakota*

*Respect.* From the very first day in class *treat students with respect,* and teach them to respect you, themselves, others, and materials in the room. This stops the need for severe discipline problems before they begin, in most cases.

Pamela A. Reed, *Grades 9 – 12, Michigan*

I strongly believe that a teacher who is prepared, motivating, and respected by his or her students will not have a discipline problem in the classroom.

Jennifer Ortman, *Grades 9 – 12, South Dakota*

Students should be taught the rules in your classroom. Personal gain is a strong motivator, so we should teach students – not tell them – the personal advantages or benefits in choosing to behave in a manner we deem appropriate. These rules must be clear and precise. Rules should be few in number and consistently enforced. Important routines should be established early in the school year, and these routines must be reinforced and maintained through the year if they are to continually be effective. Rules should and must fit both the situation and the environment.

Ron Adcock, *Grades 9 – 12, Iowa*

Be consistent. Establish rules of behavior at the beginning of the school year, either by posting classroom rules or providing them on a handout. Never say you are going to do something such as send a student to the office if you do not mean it. Follow through on such ultimatums, or do not make them in the first place.

Barbara B. Bryant, *Grades 8 – 12, Alabama*

At the beginning of each year I inform each of my classes that I expect each of them to behave as ladies and gentlemen, and that each of them will be respected as an individual, just as I expect them to respect me as an individual. Showing each student that he or she is a respected, valuable person helps bring out the best behavior. If disciplinary action is needed, it should be fair, and the incident should be forgotten. A teacher should always be professional.

Howard L. Meserve, *Grades 11 – 12, New Mexico*

---

When I started teaching school, I wanted to be friends with all my students. I wanted them to like me and I wanted to like them. Through the years, my feelings for the students have remained, but I did learn that, even if we were friends, I needed to demand respect.

At the beginning of the school year I list a few rules and regulations. My most important rule is: show respect. The quality covers a large spectrum. I tell the teens that I demand as much respect as I am willing to give to them. If my students respect me, themselves and their peers, we will all have fewer problems. I ask very few students to leave the room because we care about each other and we can laugh at ourselves.

I do not expect my students to produce anything that I am not willing to produce. I expect my teens to work as hard as I do, and hope that they will appreciate the work ethic. I ask that they try just a little harder and work a little more than the average person.

Marilynn Sexton, *Grades 9 – 12, Montana*

---

In dealing with classroom discipline it is extremely important to let students know immediately what is expected, and to be firm and consistent. On the first day I meet with my students we go over a handout entitled *Course Expectations*. This document is very specific, letting them know that I am here to teach and to help them, but in order for that to happen there are guidelines that must be followed.

Elaine W. Grant, *Grades 9 – 12, Maine*

---

Be flexible and use a few sensible rules. Base these rules on what is tolerable for you, the teacher. Too many rules, too much enforcement, is too tiring. Example: Do not fight World War II over a missing pen. Find, borrow, provide another pen and keep on going.

Myra M. Gottsche, *Grades 11 – 12, Mississippi*

---

Idle hands and minds are *indeed* the devil's playground. I've found discipline problems become almost obsolete when I present the material in an interesting manner, i.e., costumes, videos, even a poem, joke, etc. Often this approach involves more work for me. For individual discipline cases, I usually have a private conference after class with the student. Should that fail, I assign after-school work alone, with me. I seldom send a discipline problem to the office, and the students seem to be proud that we handle our own problems in-house.

Thomas D. Pennington, *Grades 6 – 12, Maine*

---

It is imperative that you establish a realistic and enforceable discipline code for your classroom. Too many *don'ts* can lead to more trouble than too few.

Frank R. Gregory, *Grades 9, 12, Delaware*

---

I find that giving students a choice in the consequences for their behavior limits the classroom time a teacher spends dealing with that specific behavior. Also, the students don't feel cornered or threatened when they are given choices in the discipline for their actions. Lee Cantor's assertive discipline methods are very helpful in giving specific methods and guidelines for any new teacher, or a teacher having problems with discipline. Also, a good rule to remember is don't generalize a punishment or behavior by including the entire class. Be specific and name names. The response is immediate.

Kyle Ann Morrow, *Grades 7 – 12, Alaska*

---

The first day students come into my classroom, I talk with them about mutual respect, and I give them some basic classroom rules. It is so important to establish this groundwork; it helps to alleviate problems later. Then, if problems arise, I try never to back myself into a corner by promising to do something I cannot carry out. Rearranging the classroom in a new seating chart often solves the problem, but whatever happens, I settle the problem that day; then the next day is a new day. Students knowing I sincerely like them helps, too. They are not as prone to misbehave with someone who looks upon them as real people with real feelings.

Nell W. Meriwether, *Grade 12, Louisiana*

---

Discipline should be established from day one. Explain in clear, concise language what you expect from each person. There are times when you will make exceptions, but remember that firmness, along with being humane, is the key.

Elizabeth Burton, *Grade 9, Tennessee*

Get to know your students. Greet them at the door and say good-bye to them, individually, at the end of the class. Move around the room and stand in close proximity to potential problem students. Have something for students to do the moment they enter the class. Try to discipline them *away* from their peers. Contact the parents *before* a major problem occurs. Have a sense of humor.

Thomas Koenigsberger, *Grades 9 – 12, Illinois*

---

On the first day of class each student should be given a list of expectations which includes classroom behaviors. There should be some which are nonnegotiable, and some which may be negotiated. The teacher should then refer to these if the need arises. Being consistent with expectations is the cornerstone of classroom control. Planning classes so that the students are active learners rather than passive listeners can help avoid problems.

Nancy D. Ruby, *Secondary, Vermont*

---

Just as first impressions are most important, so are the first few days of a class in terms of setting the tone for classroom discipline. Your expectations should be clearly understood. I have my students write down my expectations and study them because the next day I give a quiz. Whatever the expectations, the motto ''firm, fair, and consistent'' is necessary.

Bob Eschmann, *Grades 9 – 12, Illinois*

---

Classroom discipline takes care of itself in most cases if the teacher has structured, motivated learning experiences for the students. Remember to be helpful and firm, and to have a consequence for inappropriate behavior that students know about the first day of class. Extreme discipline problems sometimes require a contract between the teacher and the student which includes parental and principal signatures as well.

Kay Neal, *Grades 9 – 12, Idaho*

---

One of the most important ways to deal with classroom discipline is to take preventive measures to reduce or eliminate it. Work on creating a comfortable, safe, and fun classroom atmosphere from day one. During the first week of the semester, start each day with introductory activities, such as asking students to write something personal about themselves on 3″ × 5″ cards. You can direct or focus this activity according to the class content. For example, in a writing class you might ask students to share two things they like and dislike about writing. Students can then introduce themselves with this information, or share it in pairs and introduce each other. A shorter version of this activity is to ask students to share a favorite movie, color, food, book, hobby, musician, etc., in a round-robin style. After one week, students will be more relaxed, less likely to pose discipline problems, and know some personal information about you and each other.

Brenda M. Naze, *Grades 9 – 12, Wisconsin*

---

Classroom discipline is achieved first with motivated students. Motivated students are found in a classroom with a teacher that enjoys being with the students and enjoys presenting the subject matter. Also, the teacher needs to know the difference between good-natured behavior and harmful behavior. Flow with the good-natured student — bringing this classroom disruption back to the classroom task. Make the student with harmful behavior own the problem, and make him/her solve the problem. The teacher should never take ownership of harmful behavior of a student.

David Nienkamp, *Grades 7, 10, 12, Nebraska*

---

I believe the key to good discipline in the classroom is to have an interesting presentation or activity each day. If we as teachers can make our subjects more appealing to our students than talking to their neighbor, we will not have many discipline problems. Good planning is essential so that the students will have little dead time or time off task.

Classroom rules should be simple and clear to everyone, and must be enforced consistently. I believe good teachers have fewer discipline problems than other teachers. Teachers having difficulty with discipline should focus on their teaching methods to improve classroom discipline. I don't think a person needs to be an ogre in the classroom to have good discipline.

Charles Daileanes, *Grade 8, New Hampshire*

---

It is important to let the students know exactly what is expected of them and what they can expect from you, the teacher. On the first day of class I give each student a course guide which details classroom rules, course requirements, grading procedure and lists of materials needed for the class. I also write the information from the course guide on the chalkboard, and post it on the wall to help reinforce what is expected. I always emphasize honesty and fairness from both the students and myself.

It is much easier to be more lenient in the classroom after you have established a rapport with your students and an atmosphere where learning can be exciting.

Kelly Schlaak, *Grades 9 – 12, Minnesota*

It's essential, particularly in an urban situation, that your kids know the boundaries of your "sandbox." If you don't clearly define your limits, they will spend amazing amounts of effort determining them on their own. Your own simple, indisputable, published set of classroom rules is essential. The kids need to know where your lines are drawn, and what the consequences will be if they choose to cross them. Most important—keep it simple. If the first and each succeeding offense is dealt with firmly and fairly as advertised, they will accept your rules reasonably quickly and stop spending time seeking your limits. Prior to implementation, review your plan with the administrator next in line (the one you will occasionally need to turn to for follow-up disciplinary action). His or her consistent support will be an integral part of showing the students the way things are going to be. If you have the opportunity, talk to veterans about this. Listen for positive guidance; ignore the complainers. Get a solid list of parent phone numbers. Call them at each and every infraction for which you are issuing punishment. It's time-consuming, but it works—parental support is great when you get it, and the word of your calls spreads quickly among the students.

Ken Clark, *Grades 11, 12, Ohio*

I believe discipline in the classroom is a by-product of good planning, a strong rapport between students and teachers, and a set of rules that are enforced consistently. Good planning not only means knowing the content of the daily lesson, it also means having a variety of activities where each student can contribute his own expertise and feel like a part of the class. To develop rapport, teachers must like kids and show students this by both verbal and nonverbal actions. Be friendly, smile, get to know your students as individuals. Tell your students how much you enjoy working with them and how much they contribute to your well-being and academic growth. Adolescents look for consistency in their lives. Other aspects of their lives seem to be changing from day to day (looks, feelings, emotions, etc.). One haven of stability can be the classroom as far as operating rules apply. Don't enforce rules one day, then change the game plan the next. Students want to know what is acceptable behavior. Deal with discipline in the classroom. Don't send students to the office all the time. The problem originates in the classroom; that's where a solution should be worked out. One way of doing this is to work out a contract for discipline. Acceptable behavior is rewarded, and unacceptable behavior is punished. This is a cooperative effort between the teacher and the disruptive student.

Robert H. Yunker, *Grades 10 – 12, Maryland*

After spending twenty-two years as a classroom potentate, I have determined that a number of things affect students' behavior. Probably the best way to deal with discipline problems is to prevent them from occurring. Over the years I have stumbled upon several methods that help to inhibit inappropriate classroom behavior. They are:

- *Set standards.* The first step in controlling the behavior of others is to make certain in your own mind exactly what you expect. Determine what behaviors you can and cannot tolerate. Remember that you cannot be effective if you are distracted or uncomfortable. Consequently you must devise your expectations accordingly. If you are distracted, for instance, by gum chewers, then ban gum chewing. Do not be alarmed that other teachers seem comfortable with that behavior. You must conduct your class in a manner that is most effective for you. Keep in mind, of course, that your classroom rules must not contradict the general regulations for students at your school, and must fall within the confines of system and state policies.

    Remember, also, that you cannot anticipate every possibility. Deal with bizarre instances individually. There's no reason, for instance, to make a rule such as "Don't throw rutabagas at the clock" based on a single occurrence. If, however, a given inappropriate behavior becomes a habit, amend the rules.
- *Be specific.* In the lower grades, particularly, it is important to avoid generalizations such as "Be sweet" or "Don't disturb others." Try to make all expectations and the consequences for violation as clear and precise as possible.
- *Be reasonable.* Remember that you are dealing with students who are not adults. Students are young and full of energy. They cannot be expected to sit rigidly still for fifty-five minutes without blinking an eye. Remember, too, that just because students are young and relatively inexperienced, they are not necessarily stupid. Make the negative consequences for violations reasonable and *always* follow through. For instance, if a student is tardy for class, remaining in the room for a minute or two after the rest of the class is dismissed may be a reasonable consequence. Whatever you do, make promises, not threats. Don't tell a student that he will be hanged at high noon for an infraction unless you have the noose ready.
- *Communicate expectations.* While students are not adults, they are people. If they know in advance what is expected of them, for the most part they will try to comply. Make sure every student knows the rules. I provide a copy of my rules for each student and spend the first class period of the school year discussing them. I have each student sign a form indicating that he has received the list of rules

(not necessarily that he agrees with them), that he understands what is expected of him, and that he is aware of the consequences should he violate one of the rules. I also have the parents sign the copy of the rules so they, too, may be aware of my expectations for their students. In addition, I post a copy of the rules in a conspicuous place on the classroom wall. Be sure to include rewards for good behavior as well. Sincere verbal praise for groups as well as individuals produces remarkable results. Other more tangible rewards are also appreciated. Set aside an extra half hour for computer games, time in the media center, a subject-related video or game, or time for leisure reading when a class performs well. A complimentary phone call or note home to a parent encourages repetition of a desired behavior.

- *Be consistent.* A former principal of mine ended every memo with this motto: "Consistency is the key to effectiveness." For years I gagged every time I came across this statement, but I have never doubted its veracity. Administer the consequence for inappropriate behavior *each time* a violation occurs. Nothing encourages experimentation more than intermittent consequences. You must, of course, allow for extreme circumstances. For instance, if fifteen of thirty students are late to class with the explanation that there was a fight in the hall during class change and they couldn't get through, it would be unfair to punish them for circumstances beyond their control. Check out their story before lowering the boom.
- *Be fair.* This is no time to play favorites. Administer the same consequences to each violator.
- *Plan.* Students have little time and less inclination to be disruptive when the teacher has planned the class activities well. Make your objectives for the day clear and precise. Have ample activities planned. Have all materials assembled, organized, and within reach. By nature, the education business is fraught with interruptions. There are intercom announcements, teachers and administrators at the door with requests, a student in the second row with a nosebleed, and any number of other disruptions that you will have to deal with during the course of any given class period. These cannot be avoided. You can, however, avoid disruptions caused by poor planning. Your enthusiasm for your subject and belief in its importance go a long way toward fostering the students' proper attitude toward class. When you make it clear that the material at hand is interesting and important, students will follow your lead.
- *Be professional.* Students learn best by example. Always remember that you are the adult in the classroom and act accordingly. Shouting at students, belittling them, and being sarcastic is not being a good

role model. Ninety-five percent of the time, when students misbehave, it is not intended as a personal affront. Johnny probably wasn't tardy or forgot his book in order to annoy you. In fact, you were probably the furthest thing from his mind. Anyone — adult or student — can be forced to be disrespectful. When a student is humiliated by a teacher, he naturally responds with unpleasant, defensive behavior. I have found that students who are treated with dignity and respect generally treat me with dignity and respect. By the same token, it is crucial that you follow your own rules. If you have deemed it necessary, for instance, that students not eat and drink in your classroom because you find that the rattle of potato chip bags and the gurgle of soft drinks are not conducive to learning, then don't sip coffee or munch on a donut during class. If you insist that students be properly prepared for class and arrive promptly, then make every effort to be on time and have all materials at hand.

- *Keep your perspective.* There will be days when nothing goes right. Don't despair. Think about what went wrong and why. Then try to correct or avoid the problems in the future. That's the great thing about teaching: you can start all over again every day. No matter what happens, remain calm. You can fall apart later when you're alone.
- *Keep a sense of humor.* Don't get so caught up in the seriousness of your work that you lose your sense of humor. Kids are fun. They often are very insightful and witty. Enjoy them.
- *Be honest.* The young are the first to detect deceit. When you make an error or don't know an answer to a question, don't try to cover up. Admit your error or your ignorance. Students not only learn that you are human, they also learn that it's okay to make mistakes.
- *Get help.* Whenever possible, handle your own minor discipline problems. Sometimes when students sense your uncertainty or insecurity, they will deliberately do things to annoy you. If, however, a serious problem arises, *do not hesitate to seek help.* If it is a minor, habitual problem, consult experienced teachers in whom you have confidence. They have probably dealt with the issue before, and can give you some pointers. If it is a serious problem, ask an administrator or a counselor for assistance immediately. They are there to help you.

*Anonymous, Grades 9 – 12, Georgia*

---

The one and only rule for classroom discipline is "Be consistent!" Let the kids know what you expect and praise them for compliance. However, when a rule is broken, they should expect you to do what you (or you and they) prescribed as the consequence of their action. And, most importantly, the consequence should be the same for everyone who initiates similar action.

Beverly-Ann Gill, *Grade 9, Rhode Island*

---

Since most discipline problems arise from someone's unmet needs, a first step is planning for instruction that meets everyone's needs — including yours. A well-organized classroom, where learning and students are the first priorities, is one key. Having each lesson's activities ready to go right away also helps. A written discipline, constructed with help from your students, if you wish, is important in helping you identify three to five simple clear rules with specific consequences. Implement your plan calmly, consistently, and conscientiously. Be flexible — it's a plan, not a strait-jacket — and keep your sense of humor.

Lynne Cullinane, *Grades 9 – 12, New Jersey*

## CHAPTER HIGHLIGHTS

Gain the respect of your students by being fair and consistent with your discipline. Be prepared for the class, and keep the students challenged.

## CHAPTER 12

# Motivating Students

MASTER motivation and you'll come as close as you are probably ever going to come to being master of the classroom. The ability to motivate students is not a biological gift. It is a skill which can be developed. But it is not an easy skill to develop, and it is one which is tested every day.

First, *you* must be motivated. Motivation is contagious. As motivation increases, the problem of classroom discipline decreases, and, conversely, as motivation decreases, the problem of classroom discipline increases.

Next, as a teacher you need to be a good coach — you need to help your students want to perform. Think back to the classrooms of your past. Remember your favorite teachers, the ones you wanted to perform for. They created an environment you wanted to be part of, and they did it, at least in part, with the motivational techniques of good coaches.

You can also motivate by using showmanship to add a little spark to your classes. Take some cues from the TV shows your students watch.

Another basic item which will help you motivate your students is utility. Make sure your students know what the knowledge is worth to them. When they ask, "Why do I need to know this?" make sure they have an answer.

Successive approximation is a useful motivational tool. Start the student with an intellectually easy — but honest — task, then move gradually to levels of higher difficulty. This approach keeps frustration low, and feelings of accomplishment high.

To motivate all students, you must remember that each student is different from every other student. Whatever path you follow to motivation, it must be personalized for each student.

And remember to make your students believe it when you say it!

Now add to the basics of motivating students with the following specifics from our veteran exemplary teachers.

Your own imagination is the first step. If you don't enjoy the lessons, it's hard to imagine that your students will. Try something new. Always be on the lookout. Read what you find on the front page of the paper to your students. Invite a guest speaker. Vary from your typical methods. Try a slide show. If you are interested in what you're doing, nine times out of ten your students will be too.

Also, be sure to do the work right along with the students. If you assign quiet time for writing, you should write. If they're working on word problems, work yours right along with them. Set a good example for a quick pace, and they'll try to keep up.

Sue Vaughn, *Grades 9 – 12, Nevada*

Motivating students requires more creativity and stamina every year. The MTV/microwave generation is easily bored. When I asked students what motivates them to learn in the school, the overwhelming majority answered, "fun." The number two answer was "wanting to please my parents and teachers."

The first five minutes of class are crucial. I grab their attention in any fast-paced, entertaining way I can – jokes, personal anecdotes, comic strips. Then I teach.

Although I don't try to teach to their ever-changing tastes, I do give them a practical reason to learn what I teach. This reason will not necessarily provide immediate gratification. Therefore, I also rely on their attachment to me. If I love what I do and show my own enthusiasm, often they'll become enthusiastic too.

Christine Robinson, *Grade 7, Minnesota*

Tap into your students' lives to make learning personal, real, and meaningful. Draw from their own experiences, values, world views, and hopes, without judging. Encourage them to jot down their preliminary ideas informally and individually, then share with partners of small groups before sharing in a whole-class discussion. More students will be involved, and the quality of the discussion will improve. Connect prior learning with new learning in ways students can grasp, and draw on what we know about learning styles to let students process learning personally, conceptually, and practically.

Lynne Cullinane, *Grades 9 – 12, New Jersey*

The most effective way to motivate students is to connect what you're teaching to their personal lives. The desire to know is inherent in every individual. Thus, the effective teacher's responsibility is to activate that desire by creating experiences to which the student will see the relevance of the concept about to be taught. This strategy will eliminate the perennial question, "Why do I need to know this?"

Terry Boyle, *Grades 10, 11, Virginia*

---

If you as a teacher believe in what you are teaching, and if you emote some enthusiasm, the students will be attracted to the subject. I have a koosh ball that I use in a game which is really a language or literature review. I teach linking verbs in a guru chant that former students come back to repeat to me! I allow students to choose teams (or I choose them selectively if some students are apt to be overlooked), to make hard questions in spelling, literature, grammar, and vocabulary, and to have a fast paced *Jeopardy* game. The prize may be points on a test or checks (Copernican money) to be used toward an auction held twice a year in our Copernican program.

Beverly-Ann Gill, *Grade 9, Rhode Island*

---

You won't be able to motivate your students if you are not motivated yourself. Be energetic and enthusiastic concerning your lessons. Have fun in the classroom. Contrary to popular belief, learning and having fun are compatible. Give personal examples. Students love to see that their teachers are human. Ask each student for a personal example. Classroom discussion becomes less threatening if everyone shares. Establish high performance expectancy rates for all your students. Most students will only perform at the level you expect them to. Encourage them with positive comments rather than discouraging them with negative ones. If they give you a less-than-satisfactory answer, praise them for what they've done and ask them to expand on certain aspects. Force them to think more. Students usually are willing to go those few extra steps if they sense continued success and not failure.

Robert H. Yunker, *Grades 10 – 12, Maryland*

---

Particularly in urban schools, motivation comes by accentuating the positives and building individual student self-confidence in problem solving. Too many of these young lives are far too negative in their own right. But there is a careful balance to strive for here. You don't want to make the task so transparently easy that the student doesn't gain a measure of genuine accomplishment, and yet it's important that each student learns that with an honest effort, he can and will succeed with you. Early positive accomplishments that can be enjoyed by the vast majority of the class set the pace and can make your class different from others. Once that self-confidence builds, then challenge it. Remember, many arrive with the mind-set that failure is a foregone conclusion. You earn your teacher pay when you get them to realize that just isn't going to be the case in your class.

Ken Clark, *Grades 11, 12, Ohio*

---

When trying to motivate students, I often ask myself, "If I were a student in this class, would I be excited to learn?" If the answer is "yes," then I feel comfortable about the way I am presenting classroom material. If the answer is "no," then I need to find an alternative teaching method.

When teaching a classroom full of students, one must remember all the different personalities that are involved. I try to cater to these different personalities by being creative with my subject presentations. One unit might involve a series of lectures, another will demand hands-on experience, and yet another may combine the two. The creative possibilities are endless if one is willing to accept the challenge.

Kelly Schlaak, *Grades 9 – 12, Minnesota*

---

I believe that motivation must be internal. We as teachers must focus our attention on finding ways for our students to be successful rather than a reason for failing them. The adage "Success breeds success" is quite true. Students' efforts will increase significantly if they know they will be rewarded for their efforts. I also try to make my students aware of the long-term ramifications of their classroom successes. Your classroom achievements of today will have a direct impact on your future.

Charles J. Daileanes, *Grade 8, New Hampshire*

---

Motivation of students is a matter of the teacher's enthusiasm. The teacher must know the subject matter thoroughly, and then present it, showing how much fun he/she is having. The teacher's enthusiasm is contagious. Be willing to admit mistakes. If a student proves a teacher made a mistake, that student is to be congratulated. The teacher shows that he/she is still learning with the students. Most of all, the teacher should be full of excitement and joy.

David Nienkamp, *Grades 7, 10, 12, Nebraska*

---

An important way to keep students motivated is to teach with variety, and in the process, address the diverse learning styles that exist in any classroom. Use cooperative learning groups in which all students are asked to take active roles, learning and solving problems with peers. Allow for individual learning time, during which you can work one-on-one with students. Large group discussion facilitated by you is important for developing speaking and listening skills, and for more verbal learners. The traditional teacher as lecturer need not be forsaken, as lectures are appropriate for teaching some content. Lastly, be sure to supplement with media resources and community guest speakers. In essence, maintain a stimulating and inviting classroom which meets the needs of all students. This is the key to keeping motivation levels high.

Brenda M. Naze, *Grades 9 – 12, Wisconsin*

---

Motivating students needs to be a full-time goal of any teacher. Varying the methodology helps. Using the same tests and same procedures year after year is totally ineffective. In language classes, some game techniques work well; interviews (even one class member interviewing another) can be adaptable to book reports, author's works, and other speech and literature-based subjects. Student written tests, as a means of review or for a writing skill, are also effective. Video taping certain activities and replaying them can also help create interest for students.

Kay Neal, *Grades 9 – 12, Idaho*

---

Most teenagers simply think about what they are going to do this weekend. I continually encourage students to think about what they want to do five, ten, twenty years from now. Also, I use examples of past students (positive and negative examples) as to what are some outcomes of present behavior. I believe each of my students are important individuals and that they can succeed in life. I then seek to have them join that bandwagon. Grades are not stressed as much as doing your best.

Bob Eschmann, *Grades 9 – 12, Illinois*

---

Relate the activity to the real world. Make certain that they are capable of accomplishing at least part of every assignment. Give them problems that make them stretch. Always show enthusiasm for what you are teaching. Bring in former students and young guest speakers that are also enthusiastic. Do as much hands-on activity as possible; don't do the same thing every day. And always encourage them individually and specifically; not just "nice work."

Thomas Koenigsberger, *Grades 9 – 12, Illinois*

Strive to make lessons meaningful as well as interesting. Give students an opportunity to participate in selection of some class activities. Friendly competitive events, lavish praise for accomplishments, and display of student's work all serve as motivating factors.

Elizabeth Burton, *Grade 9, Tennessee*

Motivating students is a real problem! First, if students are asked to do anything via written work, it is either checked or graded; next, students are simply *expected* to do their work. I often begin the class by reminding those who have neglected to bring in their assignments they are due. One student said, "She makes us feel so guilty when we don't bring in our work that we just do it." I don't have complete success, but by confronting the students about their assignments, they all know I am interested in what they're doing. My being interested helps.

Nell W. Meriwether, *Grade 12, Louisiana*

Setting up an assignment that everyone can be successful at is a wonderful motivator. For example, at our school we have a computer lab with sixteen computers. After assigning a research report to my seventh grade English students, we did research in the library. Then we went to the computer lab where each student wrote his report on his own computer. As they were writing, I went around the room to help edit each report. I also helped edit their footnote and bibliography page. Editing on the computer was easy and painless. Consequently, all of my students received A's for this assignment. Therefore, every student felt proud and successful. Each one has a perfect example to use for future reference. Also, those that finished or were waiting for my help had fun doing graphics and playing with different formats and fonts for their title page.

My second recommendation is for the teacher to enable any student who is not satisfied with a grade on an assignment, quiz, or test, to retake or redo it. The teacher should definitely encourage the students that received a failing grade to redo or retake the assignment. Eventually, failing students become self-motivated to succeed the first time.

Kyle Ann Morrow, *Grades 7 – 12, Alaska*

You must find a way to make each student feel you are genuinely concerned with him/her. Private talks with the student, his/her coaches, or parents are sure ways to demonstrate that you care. If the student believes that you care he/she will work for you. They do not want to let you down. Also, incentives work—e.g., grades and names posted, smiley faces, etc.

Frank M. Gregory, *Grades 9, 12, Delaware*

---

Smile. Be interested in your subject(s) and know about the interests of the students that might relate to it. Let them participate with each other, with you, and in front of their classmates. Participate with them, lead them toward action. Ask questions, do not be discouraged. Smile a lot.

Myra M. Gottsche, *Grades 11, 12, Mississippi*

---

When teachers like what they are doing, students become motivated. They want to join the action. I love my job. The greatest satisfaction comes after I have worked with a student for a while and then can see how that student has grown as a writer. I am interested in my curriculum, and I try to make it interesting. With the age of technology, there is always something new to learn. Education is challenging. The students catch the fever from the teacher. Learning becomes contagious.

Students will work if the atmosphere is pleasant and the workplace is exciting. Teachers need to encourage students constantly. Teens need the constant interest and approval of the teacher. I try to have fun with the students and the curriculum by relating the subject, whatever it might be, to the present time. If I am teaching *Romeo and Juliet,* I compare Shakespeare's young lovers and their woes to the situations that teens may be experiencing. If I teach publications, I bring in all kinds of magazines and newspapers so students can see models and try to do the same kind of work. Sometimes I have to admit that the work is not so exciting. If I admit that I don't really enjoy an experience, the students relate. They learn that they experiment with all kinds of experiences in the realm of learning.

Marilynn Sexton, *Grades 9 – 12, Montana*

---

*Promising* is a motivator—"Will you promise to get this done if I let you work together?" The key is to get the promise first. Then you can integrate cooperative learning with this method. An upcoming event can be motivational. The upcoming activity can activate. The event may be a social event at school, the next big test, a reward for finishing, or no work for the weekend because work is finished. Make the event seem like a benefit, not a threat. A technique that is used all too often is to give instructions and walk away assuming students will do as directed. It won't necessarily happen!

Ron Adcock, *Grades 9 – 12, Iowa*

Students need to experience success. From infancy on, we all want to repeat a performance that gains approval from our peers. If a student gets half a sentence correct in my foreign language class I praise them for that fifty percent. "You did a great job with the noun and verb. Now let's work on making the adjective agree with the rest of the sentence." If the whole sentence is wrong, I tell them I appreciate how they speak out nicely or pronounce the words correctly. It's important to me that students feel "safe" in my classroom. I want them to know they aren't making mistakes, but are a part of a learning process.

Jennifer Ortman, *Grades 9 – 12, South Dakota*

---

Relate materials to the students' past, present, and future. Show students how and why they need to learn the material presented. It is very necessary to use instructional methods and information that the students value.

Pamela A. Reed, *Grades 9 – 12, Michigan*

---

Be enthusiastic! If you are not excited about what you are doing, students certainly will not be.

Sally Tanner, *Grades 10 – 12, North Dakota*

---

Administer a diagnostic test at the beginning of the year to determine the level at which you should teach. Teach at students' level of comprehension and relate the subject matter to students' everyday lives and what they might gain from the material in the future. Avoid long lectures; instead, create fun activities to teach each lesson at higher-level thinking. Ask a lot of questions and avoid giving direct answers to questions. Instead, ask questions that would lead students to the answers. Allow all your students to taste success; then, gradually raise your standards and expectations. Provide extra opportunities for students who fail, and incentives for those who excel.

Abbas Saberali, *Grades 7 – 9, Texas*

---

Enthusiasm is contagious; be enthusiastic every day! Success breeds success! Create opportunities for success, and then praise! Give only relevant assignments. Be creative; vary your approach. Reward listening and learning from mistakes. Provide prompt, courteous consequences for all efforts (or lack of them). Tell students exactly what you expect and keep your word.

Elizabeth Bishop, *Grades 9 – 12, Michigan*

---

Sincerely *care* about your students. Let them know your enthusiasm and love of the material and interest in them. They can spot a phony, so be honest. If there is a unit you have trouble with, let them know that.

Susan LaPeer, *Grades 9 – 12, Massachusetts*

---

Give stickers for A papers. Display papers with stickers on the bulletin board. Have a student-of-the-week posterboard display for the student who has earned the most A's during the past week or the student who is most improved. Reward the student-of-the-week with special privileges, like a library pass, etc.

Brenda Clemons, *Grades 10 – 12, Virginia*

---

I encourage the students to speak Spanish by allowing no English to be spoken in my class. I use a total immersion approach and emphasize conversation. The student learns by continual practice with the language. If I see any of my students outside the classroom I speak to them in Spanish and reward them with extra credit if they converse back with me in Spanish. We also travel to Mexico annually. They must speak Spanish during the week, which is a good incentive for them to develop their skills.

Paul C. Worthen, *Grades 9 – 12, Arizona*

---

Students will be motivated by the knowledge that what they are learning is something they truly need to know. Have discussions with your students about *why* they are doing certain assignments. Make learning as relevant and as meaningful as you can. Remember this, too — people don't care what you know until they know that you care. Care for your students. They'll know if you really do or not.

Kimberly Sturgeon, *Grades 9 – 12, Virginia*

---

Stress intrinsic interest of topics, importance to lifetime goals. Develop a two-tier testing procedure; first, a standard procedure test, second, a classroom review with students copying the correct solutions to replace the usual test. The result is better classroom morale, and better attentiveness to the subject.

David Hungerford, *Grades 9 – 12, New Jersey*

---

Awards and certificates prominently displayed for all to see are highly motivational for high school students. In addition, I allow students who make an A on a test to choose one free assignment during the grading period. They receive full credit for that assignment. A postcard sent home indicating outstanding achievement is another method that has been very successful.

Gayle Blunier, *Grades 10 – 12, Indiana*

---

I give my motivational speech on day one. We discuss reasons for students to do their best, such as graduation requirements, class ranking for scholarships, ACT tests, job placement tests for hiring. I expect these traits from them – maturity, dependability, responsibility, involvement, priorities, and balance. The teacher must show enthusiasm for the daily lessons, and a little showmanship never hurts! If the students see that the teacher is excited, they become interested. I also tell my students (and encourage them) to ask questions. I often say ''That's the best question I've had all day.'' Praise them individually and as a class.

June Enowski, *Grades 9 – 12, Missouri*

## CHAPTER HIGHLIGHTS

To motivate students, your lessons should be meaningful, useful, relevant, and personalized. Give the students a reason to learn.

## CHAPTER 13

# Relations with Parents

IT is interesting to note that parent relations ranked as the top challenge at the elementary level, but dropped to number three at the secondary level. Elementary parents may have more of an impact on their children's education than do high school parents. While not the most important challenge for high school teachers, parent relations are still important.

As a beginning secondary teacher, you can involve parents in their children's education, but it takes charisma, sincerity and tenacity. Parents must be invited to be — and accepted as — full members of the parent-teacher team. You must make the first move to establish good rapport. After you establish rapport, focus on the most important reason for good parent-teacher relations — keeping parents informed about their children's progress, your classroom, your expectations and about the school in general.

---

In urban schools, establishing parental contact is important and takes initiative. Getting good phone numbers early in the year is vital. My students each fill out a fairly detailed autobiographical sheet during the first few days of class. This way I gain insight into current family arrangements, and residential and business phone numbers in the process. Build a notebook with a page for each student, with parent names and numbers at the top to keep a record of calls. Use an occasional prep period (if you have the luxury) to establish contacts early in the year. Then call — don't put it off. Have something positive to say, and let the parent know when you can welcome return calls or visits. Also, if a student crosses one of your boundaries requiring any form of punishment, call the parent and work to solve the problem with them. When you do this, you may not always get instant results with the student, but your overall classroom discipline success will increase dramatically as the word spreads among the students that you care enough about situations to call home.

Ken Clark, *Grades 11, 12, Ohio*

Communicate, communicate, communicate! Send a letter to each parent before the beginning of the new school year asking the parent to write back about the special qualities of your student-to-be. Write to the students to welcome them to your class. Involve parents in classroom projects by engaging them in curricular activities. Send a note to or call every parent at least once each year with one bit of good news about his or her child. When there are problems, focus on the problem (not the child) and seek the parent's help in problem solving. Put the emphasis always on working cooperatively together to insure the child's success.

Lynne Cullinane, *Grades 9 – 12, New Jersey*

---

After the first homework assignment of the year, call every parent whose child didn't turn in the work. Be persistent! You won't believe the response. Parents are much more receptive to corrective measures at the beginning of the year than they are to those at the end.

Don't be wishy-washy in a conference about a problem student. If the kid is punching people around him/her, don't say "He seems to have communication problems." Don't say "She has difficulty getting her homework in on time," say "Diane hasn't turned in seven of the last nine assignments." Specificity helps. Don't make a problem worse or better than it really is—just provide as much information as you can.

Sue Vaughn, *Grades 9 – 12, Nevada*

---

By all means, get parents involved! Be honest with them about their child's progress in your class. Remember, if you must say something derogatory about the child, preface it with something nice! Remind them that you may need their help, and you may be calling on them if you need them. Recently, one parent got so involved in our Shakespearean festival that she wanted to sew a costume for herself so that she could attend! It took a great deal of diplomacy to show her how setting such a precedent could involve eighty-seven other parents with forty-four kids in a small space! We made a video of the event that she was allowed to borrow.

Beverly-Ann Gill, *Grade 9, Rhode Island*

---

For too long, parents have been seen as teachers' adversaries. We must place parents in a more positive situation. We must make them our allies. Historically, poor student performance has been reported to parents by phone messages, signing of the bad test grades, etc. I've tried to turn this concept around by contacting parents whenever my students do any kind of positive work. By teaching in a small rural school, I come in contact with a lot of parents outside the school setting (supermarket, bank, etc.). Take a little time to talk with them about their children's progress. This demonstrates a continued interest in their children's well-being. Become a community spokesperson for your school. I send messages to various service organizations, volunteering to be a guest speaker. This will tie the school and community closer together. Invite parents to come and sit in on your classes. Parents can get a firsthand view of what is going on in education. I also send home information as to how parents might help their children study and prepare for tests. Make the parents an integral part of the learning process.

Robert H. Yunker, *Grades 10 – 12, Maryland*

---

Openness and honesty are essential to a good relationship with parents. If parents know you are sincere in your efforts to make their child a better person, you will have very few problems.

My grades are computerized and the students know their averages at all times. There are no surprises. I also save all the work that students do in my class. This gives me visual proof, if necessary.

All communication should be prompt and courteous, even if it is trying at times. We are professionals and should always act in that manner.

Charles Daileanes, *Grade 8, New Hampshire*

---

Visit with parents whenever possible about their children, their school, their community, their job, etc. Take advantage of the obvious – parent-teacher conferences, PTA, open-house night at school, whatever the school does to get parents and teachers together. And go beyond. Visit with parents at ball games, church, the cafe, etc. Let them know you.

David Nienkamp, *Grades 7, 10, 12, Nebraska*

---

Communicate with your parents often. Call them for positive as well as negative developments. A short note is also very effective – don't be adversarial. If there is a potential for a serious problem, make certain a third person sits in on conferences, e.g., principal, department head, counselor, etc. Get them involved in school activities and thank them for their help.

Thomas Koenigsberger, *Grades 9 – 12, Illinois*

---

Make parents realize they are an important part of their child's school work and of the overall school program. Keep an open line of communication between teacher, child and parents. Phone calls, chaperoning activities, room parents, and classroom visitations aid in good parent relations.

Elizabeth Burton, *Grade 9, Tennessee*

---

Always remember, the parents pay the bills. Send home reports to the parents when the student is doing a good job as well as a not-so-good job. Be upbeat and positive when talking about their child. It is very successful when you can tell a parent that there is hope for their child to achieve what they *realistically* expect.

Frank R. Gregory, *Grades 9, 12, Delaware*

---

Try to remember parents are tense with teachers. Try telephone conversations first. Be positive about their child if only in vague terms.

Example: ''This was one of our better days'' (compared to what or whom is unimportant).

You may need the parent one day. Begin cooperation before you have to demand it.

Myra M. Gottsche, *Grades 11, 12, Mississippi*

---

Be totally ready for a parent meeting. Don't wait until just before to organize your thoughts. Check with administrators and counselors; they might have valuable information you will need for the meeting. Be professional. Do not make parents wait. A warm, not cold, matter-of-fact attitude will help ensure their trust and confidence. Shake hands with both parents. Keep your emotions under control. Make sure you listen to parents before you say a word. Don't argue. Parents are seeking your support, and you're the expert. When parents are finished, be honest but gentle. Don't build false hope, present your viewpoint truthfully, don't try to deceive.

Ron Adcock, *Grades 9 – 12, Iowa*

---

Always start the conversation with a positive comment about their child. Be fair in your presentation of the problem, but *be honest.* End the discussion with a positive and helpful answer to the problem.

Pamela A. Reed, *Grades 9 – 12, Michigan*

---

Be warm and outgoing. Get in the habit of phoning or writing to parents *before* there is a problem in class.

Sally Tanner, *Grades 10 – 12, North Dakota*

Confer with parents of your students as often as possible. Parental involvement is the key to students' success. Student behavior and performance in the classroom cannot be separated from conduct at home or elsewhere; therefore, there must be an open communication line between the teacher and the parents. Prevent discipline problems and failures by identifying students that are heading in that direction early on and getting their parents and counselors involved in the process of counseling and motivation.

Abbas Saberali, *Grades 7 – 9, Texas*

Send home good news often; praise and attention will reap great results! Persistent problems that could lead to low grades or major disciplinary action must be reported. Demonstrate a sincere interest in their child's achievement and well-being. If you have earned their respect, they will support your efforts—and that is essential. Document all contacts. Report cards should not contain surprises!

Elizabeth Bishop, *Grades 9 – 12, Michigan*

Communicate *positive* sentiments even if you have trouble with a student. Genuine concern and some positive statements go a long way.

Susan LaPeer, *Grades 9 – 12, Massachusetts*

Contact the home with good news as often as possible. Write a brief note stating that "your son/daughter had a very good week" and send it home with the child. Add a sentence or two to make it specific: "He raised his hand before answering questions" or "He stayed in his seat." When a problem arises that needs parental intervention, feel free to call on them immediately. You have already established a pleasant relationship with them through your notes.

Brenda Clemons, *Grades 10 – 12, Virginia*

I strive to involve parents as much as possible. Extra credit is given to students whose parents attend parent-teacher conferences. I encourage the student to participate in these conferences, too. This gives me the opportunity to show the parent firsthand the level of competency attained by the student. It also allows good interaction between the three of us. If there is a concern, it is beneficial to have all parties discuss it together. My class is always open to visits by parents. If I have a concern during the year, I try to make the parents aware of it if I think they could help.

Paul C. Worthen, *Grades 9 – 12, Arizona*

---

Generally, I find parents quite responsive to any request to help with their children's classroom problems. The problems that one encounters with parents are the overwhelming ones that they themselves suffer from as victims; for instance, racism, poverty, and single parenthood.

David Hungerford, *Grades 9 – 12, New Jersey*

---

Sometimes educating the parent is one of the first steps in educating the child. Being tactful, yet honest, with parents is essential for a good relationship. I approach a meeting involving parents with a positive attitude and usually mention early in the conversation something like, ''What can *we* do to help Mary?''

I always refer to the child by name in the discussion and avoid talking about someone else's child. It's essential to remember that even during a tense conversation, the parent(s) and the educator are on the same team.

Thomas D. Pennington, *Grades 6 – 12, Maine*

---

I try to get to know as many of the parents as I can as soon as possible. I let them know what we will be attempting to accomplish in the class, and how their child's work will be evaluated. I also attempt to learn if there is any special information I should know about their child (allergies, etc.). Establishing a relationship early leads to better rapport if problems arise later.

Howard L. Meserve, *Grades 11, 12, New Mexico*

---

In dealing with parents I have found that being up-front and honest works best. I ask my students to take their *Course Expectations* handout home to share with their parents. This informs the parents immediately of what is expected, and answers many of their questions regarding grading, homework, etc. It is also important to keep parents informed of their child's progress, which I do by preparing a progress report for each student at midquarter.

Elaine W. Grant, *Grades 9 – 12, Maine*

---

Recognize that parents have a vested interest in their child's education and that they can be a powerful friend to the classroom teacher. One must be aware that many parents still feel like a child in the presence of a teacher. Be careful not to intimidate or preach to them when talking about their child. Don't hesitate to call parents. They want to know if their child is having problems. Invite them to be active participants in their child's education.

Nancy D. Ruby, *Grades 9 – 12, Vermont*

---

He/she is their baby. They have never seen the little hellion that you are describing. Exercise all possible patience and understanding, but don't sugarcoat the problem. Try to remember, too, that in their eyes you represent the school. Promote a positive image.

- *Be agreeable.* Regardless of how angry you may be with a student, try to be agreeable and put the parents at ease. A minute or two of pleasantries won't hurt. Always introduce yourself and state your position.
- *Be calm.* When dealing with parents, just as when dealing with students, try to remain calm. Again, the parents aren't really angry with you, but with the situation.
- *Listen to parents.* Allow the parents to express their concerns first, particularly if they have requested the conference. Knowing exactly what the issues are will eliminate wasted time. Then address those issues specifically. Then mention any concerns you have that have not been addressed. If at all possible, ask the nature of the visit when the appointment is made so you will have the necessary materials on hand for the conference.
- *State facts.* If there is a problem, then state it in plain terms. Pretending it doesn't exist, or minimizing it, won't result in any changes. If the student has failed to bring in seven out of ten homework assignments, then say so. Be specific. Your conduct record is vital here. You can say, ''Fred has fallen asleep in my class six times in the last two weeks (give dates if need be). I am concerned that he is not getting enough sleep at night. Can you shed any light on this problem?'' instead of ''He sleeps in class.'' Don't use vague terms like *sometimes* or *seldom.*
- *Provide documentation.* Take your attendance records, conduct records, grade book, samples of the student's work, etc. with you to the conference. When a parent sees the actual test with the poor grade or the attendance sheet with the nine absences marked, your case is far more credible.
- *Avoid making judgments.* Deal strictly with observable behaviors. Concepts such as *a bad attitude* or *lazy* condemn the whole person. Your concern is with his behavior and academic progress in your class.

- *Provide positive communiques.* Take the time for a complimentary note or phone call when a student performs well, or shows notable improvement. A parent who has heard from a teacher when there has been smooth sailing is more receptive when problems arise.

*Anonymous, Grades 9 – 12, Georgia*

---

The first thing every teacher should remember is to treat each parent's child as if he or she is the only student you have to deal with. Johnny may be one of 130 for the teacher, but to that parent he may be one of two. Every parent wants to feel as if his child is important to this teacher. The first six weeks of school each parent should be contacted by a teacher. If you work on a team, each teacher could call his homeroom. Explain the team policy on grades and discipline and answer any questions about the policies. A little time at the beginning of school can solve a lot of problems later in the year. Of course, if the teacher is self-contained, the teacher should contact every parent.

*Elaine Puckett, Grade 8, Georgia*

## CHAPTER HIGHLIGHTS

Build a relationship with each parent through written and verbal communications. Be sure to report positive happenings as well as negative ones. Always show concern for the student.

## CHAPTER 14

# Classroom Management

ELEMENTARY and secondary teachers both ranked classroom management fourth on their list of challenges. Good classroom management alone won't make you a successful teacher, but it will improve the quality of your daily life. Classroom management is definitely related to discipline.

Efficient use of your time is the bottom line of classroom management. You probably need to add a little extra time on both ends of your day. The extra time at the beginning of the day is probably most important; it gives you a chance to put the final touches on your plan for the day. Some teachers employ a very structured or formal daily routine. Others are more flexible. What really matters is that you do, in fact, have a plan.

Rarely will all things go precisely as you had them planned. Too many things are out of your control. Be flexible in your daily plan, expect to have to use your flex, and, most importantly, have a reasonable number of appropriate activities in reserve.

It's also important to vary your routines. Doing the same thing day in and day out becomes boring. When bored students seek other forms of stimuli, all too often they are associated with misconduct.

One last related item is the need to keep your students on task—appropriately. Note that we do not necessarily suggest *busy* or *occupied*. Don't give your students meaningless things to do just to keep them busy. Keep your students on task, because if you don't they will take the initiative to activate in other areas, probably not consistent with your overall plan for the day.

Here are some classroom management techniques from our veteran teachers.

---

Allow a few minutes at the beginning of the period for roll call, handing out papers, etc. Spend approximately twenty minutes explaining new concepts or lecturing. Allow the remainder of the period for hands-on work, such as completing worksheets, or reading and answering questions from the textbook. Breaking the monotony of doing one activity for fifty minutes will help keep students on task.

Barbara B. Bryant, *Grades 8 – 12, Alabama*

---

I begin each day establishing the anticipatory set which, for my subject, algebra, is reviewing what we did yesterday. Sometimes I send students to the board to work a problem from yesterday. They ask questions about their assignment, which I or other students answer, or I lead them through self-discovery. Papers are then checked in class or collected for me to grade. Then I state today's objective, supply the instructional input, model the ideal behavior, and send students to the board again to check for comprehension. Then we do guided practice by starting their assignment for the next day together. There is usually a little time left for independent practice and a wrap-up of what we learned that day.

*Anonymous, Grades 9 – 12, Missouri*

---

I arrive at school at least thirty minutes, usually forty-five minutes, before my first assignment. I am able, therefore, to begin each day with an unhurried feeling, and am not duplicating worksheets or tests at the last minute.

Gayle Blunier, *Grades 10 – 12, Indiana*

---

When students come into the room, they see on the board the objectives for the day. There are always three, because students need a change of pace every fifteen minutes or so. They are given a handout each Friday with their week's assignments. Everything is ready for the day – the overhead projector has the transparency on it, or the videotape is ready, or the xeroxed handouts are ready. In addition, each class has a box for their work to be put in, which also is used for work to be given back to them. The key to classroom management is *organization* – planning your work and working your plan.

Nell W. Meriwether, *Grade 12, Louisiana*

---

I have the same room for all my classes. I see my classroom as my home. Therefore, I want it to have a very pleasant atmosphere. That means neat, colorful, and orderly. I have maps, pictures, and news articles that reflect the whole world. At the beginning of each day we discuss current news – local, national, and international. When homework is assigned, the students have the option of looking at various magazines and books I've brought to class, but their homework must be turned in later.

Bob Eschmann, *Grades 9 – 12, Illinois*

---

Always have a total lesson plan for each subject, each day of the week. This does not mean that you cannot deviate from the plan. It is easy to deviate because of changes in the day beyond your control or the gut feeling that you may have about something else working better than your original plan. It is not easy to pull something out of the sky if you have not planned at all. Write each class's assignments on the board each day.

Kay Neal, *Grades 9 – 12, Idaho*

---

Upon entering my classroom each morning, I head straight for the student desks, arranging them as needed for the first hour class. On my desk I keep my lesson plan book, in which, during my prep time or after school, I have sketched out brief outlines of what I plan to cover each hour. In trays on the file cabinet next to my desk are the materials needed for each of my five classes. On my desk are two file folders. One is labeled: *to be copied,* the other is labeled: *to be corrected.* Throughout the day, as I collect student work or have a moment to be thinking ahead, I insert materials into these folders. These few steps help make my fifty minutes of preparation more efficient, and keep me organized throughout the day, week, and year.

Brenda M. Naze, *Grades 9 – 12, Wisconsin*

---

The best advice for this problem is to be prepared for each class. Have all materials ready and, if necessary, laid out, when the students arrive.

I use a lot of group learning situations in the classroom. It is an effective way to present subject matter, but it does entail extra work on my part to get the materials ready. On days when using this method, I have student desks arranged and all materials in place before the students arrive. It is important to begin the class presentation immediately. This is when the students are most attentive and it avoids losing their interest to other distractions.

Kelly Schlaak, *Grades 9 – 12, Minnesota*

---

The way I arrange my classroom is one of my most effective classroom management tools. This preventive discipline work happens even before students arrive.

My goal is to provide as little distraction as possible. I arrange desks in long rather than wide rows. Aisles are as wide as possible. My desk faces theirs. Kleenex, trash cans, and textbooks are at both ends of the room.

Christine Robinson, *Grade 7, Minnesota*

---

In the urban classroom, students seem to relate best to the themes of organization and predictability. My lesson plans are quite detailed, and look ahead one complete week in advance. This is done to provide a set series of approaches or fronts to a particular learning objective. In these plans, I use three distinct learning scenarios, and don't introduce others without a sound explanation first. Each method calls for different but easily understood behavior patterns from the students. They know in advance what will be expected in terms of participation and behavior. That my three forms are lecture/teacher demonstration, group problem solving, and laboratory situations is not important. What is important is that I work in a framework that is conducive to effective planning, and in one that the students are comfortable with and well adjusted to. Students respond well to predictable order. The life of the urban youth is often chaotic at best, and this approach provides a measure of needed stability.

Ken Clark, *Grades 11, 12, Ohio*

---

I have been organizing my units into notebooks for the past several years, and have found that this works better than anything I've ever tried. All the materials, vocabulary, visual aids, lecture notes, project ideas, tests, texts and everything else are put into large three-ring notebooks. They are easily transportable, easy to add to, and convenient to lend. I never lend just an exercise—I loan out my whole notebook. Not a soul I've loaned to has ever returned a notebook in less than perfect condition.

Sue Vaughn, *Grades 9 – 12, Nevada*

---

If possible, give students their assignments a week in advance. Put the day's assignment on the board. Make certain you have done all the problems yourself so you are aware of potential questions and difficulties. Set down expectations at the beginning of the year and stick with them. Always have something for the students to do when they enter the class. Change seating arrangements periodically. Use at least some cooperative learning. Always have materials ready to go before class starts.

Thomas Koenigsberger, *Grades 9 – 12, Illinois*

---

Take care of routine duties in a manner that does not consume a lot of time, e.g., silent roll call while students are reviewing for a quiz. Plan lessons in advance; on occasion, allow students to assist with checking papers; vary teaching procedures and evaluative methods; and, always have an emergency plan.

*Elizabeth Burton, Grade 9, Tennessee*

---

Recognize your limitations. Plan to limit confusion. Balance your day with activities interspersed with quiet time. Everything takes longer than you think, so you cannot do it all. Recognize your energy limitations.

A day with ten objectives and ten activities will be remembered as a day of confusion by the students, and trial by fire by the teacher. Less is better.

*Myra M. Gottsche, Grades 11, 12, Mississippi*

---

Keep folders with students' names on them. Students keep old and current assignments inside. This helps eliminate lost assignments, aids in keeping track of grades, and helps students in studying for tests.

*Pamela A. Reed, Grades 9 – 12, Michigan*

---

Prepare materials and activities needed to teach each lesson in advance. Know exactly what the students are going to learn and how you are going to teach it. Establish routines to save time and avoid confusion. Assign duties to students for distributing and collecting materials. Require students to carry a binder to keep their work organized, and provide them with an assignment sheet to keep a record of class assignments and activities.

*Abbas Saberali, Grades 7 – 9, Texas*

---

Use a seating chart. A small, review warm-up assignment should be on the board daily when students enter. Soon they will be punctual, prepared, and productive while you take attendance. Wander the room while teaching so everyone has a chance to be in the front row! Formally end each class to signal proper quitting time. Never sit at your desk; use a podium and tall stool. Keep students busy!

*Elizabeth Bishop, Grades 9 – 12, Michigan*

---

I have found that the most effective organization of class time for me as a high school mathematics teacher is the following structure:

- correct homework and answer questions
- collect homework, pass back previous assignment, and do sponge activities

- daily lesson (lecture, cooperative learning, guided practice, etc.)
- independent practice

Being organized and well prepared is crucial in making effective use of class time.

Elaine W. Grant, Grades 9 – 12, Maine

---

Know exactly what you are going to do during the class. Clearly written plans and objectives are a must, regardless of how long you have been teaching. Plan classes so there are a variety of methods used. Check frequently to determine whether or not your students can demonstrate that they have learned what your objectives say they will have learned. Never try to wing it.

Nancy D. Ruby, Secondary, Vermont

---

Since we all have different temperaments and tolerance levels, keeping things running smoothly is a skill learned primarily through experience. Here are a few ideas that I have gleaned over the years that have served me well. You may find them useful. You may not. Experiment until you find what works best for you.

- Use looseleaf notebooks. In my mind, the looseleaf notebook is the most serviceable invention in the world of education. Looseleaf notebooks allow materials to be added, subtracted, rearranged, removed for copying and returned. They come in a variety of sizes and colors and have as many uses as penicillin. I use several for various purposes. Here is a sampling of their uses:

  *Attendance.* Documentation and accountability have become the teacher's bywords. Teachers are responsible for the welfare of the students in their classrooms. To be certain that I can account for each student's whereabouts during my class period, I keep an attendance notebook. The notebook is divided by tabs into five sections, one for each class. Within each class section I keep attendance sheets arranged in alphabetical order by students' last names. Each sheet is divided into two sections—one for each semester. On the top of the sheet I note data concerning the student, including his student number, book number, date entering the class/leaving the class, etc. Then as I flip through the pages calling roll each class period, I note the date on the sheets of those students who are absent. If a student arrives after roll call, I note the time of arrival in the tardy column. When lack of attendance becomes a problem, I have all the information at hand to discuss the situation with the student, his parent, or a counselor.

On the back of the sheet I note the parent's name (so frequently it is different from the student's last name) and telephone number. I also keep a record of the student's conduct on the back of the attendance sheet, as well as my reactions. This helps me to note when a student's behavior begins to become a problem and take measures to eliminate it. For instance, I note when a student fails to bring in homework on time, has to leave the room to go to the rest room, falls asleep in class, etc. Of course anyone from time to time has to be excused to the rest room or forgets his homework, but when these things become excessive, it is time to discuss it, first with the student then with a counselor or parent, to try to find a solution to the problem.

*Make-up work.* With five different classes and as many preparations, it is difficult to remember who was absent last hour, much less what specific work a student needs to complete for make-up work from last Tuesday. For each class, I keep an assignment notebook. At the end of each day, I make a simple note of work done in class, and any work that needs to be turned in. It is important to record the day's work after the fact. Too often things do not go as planned. I try to keep it simple.

I make it one of my rules that a student is to check the assignment notebook when returning from an absence and complete the work within three school days. I include in the notebook copies of handouts that the student may need. This method puts the responsibility for completing make-up work on the student, thus allowing you more time for other activities.

*Preparation notebooks.* For each class I keep a large looseleaf notebook divided by tabs into units. For each unit I keep my own notes regarding the topic. In addition, I keep a list of activities appropriate to the subject, a copy of any handouts and tests used, a list of available audiovisual aids, etc. I indicate on each activity whether it worked well or not and for which type of class: basic, general, or advanced. If I try something several times and am not happy with the results, I discard it. If I come across something new, I add it. After teaching that course for a couple of years, I accumulate a wide variety of materials which are appropriate. While initially this is time-consuming, in the long run I have found it to be most helpful.

- Delegate tasks. There are several routine tasks that even the students can help you do. For instance, I place all graded, ready-to-return papers on a certain shelf in the room. I place a rubber band around each stack and label it with the class period. In each class I designate a reliable student to return papers. Upon entering the room, he automatically picks up the stack and returns the papers as I call roll. In addition, I designate another student (one with neat penmanship) to record the day's activities in the assignment notebook during the last two or three minutes of class. While it is not wise to have students grade papers or do other tasks that involve evaluation of other students, there are any number of things that students can and will do. Try to rotate the tasks so that everyone has an opportunity to contribute. This method not only helps you with routine chores, it helps foster a sense of responsibility in the students as well.

- Designate areas. After arranging the room to allow for easy movement, designate certain areas for frequently used items. For instance, placing the trash can beside the door in plain sight and leaving it there allows students to deposit trash as they enter and exit the room rather than wandering around looking for it or leaving the trash on the floor. Have a place for students to put their papers. Years ago I purchased a plastic, paper-sized basket (it happens to be pink). I leave the basket on a specific shelf. When students complete work, they place the papers in the pink basket. This way, I am not interrupted by students asking what to do with their work. I collect the papers from the basket when I am ready to deal with them. That way none of the papers get lost. In addition I have another basket labeled *papers without names: no name—no credit.* Rather than spending time trying to figure out to whom a paper without a name belongs, I drop it in that basket after grading it. When a student's paper is not returned, he knows to check the no-name basket, put his name on it, and put it back into the pink basket so the grade may be recorded.

- Post routine. On my chalkboard I post and update daily routine information.

     *Date and objectives.* On one side of the board I post the date. Under that, I list the class period and a word or two to identify the material that is to be covered that day. This helps to keep both me and the students focused on the task at hand.

     *Initial task.* Providing a simple, routine task for the class to do immediately upon entering the classroom helps to settle the students down and put them in a working frame of mind. As English teacher, I emphasize vocabulary development. Each week the class has a sentence or two on the board utilizing one or two of the vocabulary words from the week's lesson, leaving a blank for words. For instance, if one of the vocabulary words for the week is *interminable,* the sentence might read: ''For the students and the teachers, the wait for spring holidays seems _____.'' The students are to copy the sentences on a daily basis and fill in the correct word from their vocabulary list. At the end of the grading period, the students turn in the sentences for a grade.

At the beginning of each unit, I post on the board by class periods a list of upcoming major tests and dates. For example, the left-hand side of my board looks something like this (first period is my planning period and therefore is omitted here):

| Hour | Test |
| --- | --- |
| 2 | Chap. 12 |
| | 3-31-92 |
| 3 | Poetry |
| | 4-6-92 |
| 4 | *Iliad* |
| | 4-4-92 |
| 5 | Poetry |
| | 4-6-92 |
| 6 | Chap. 12 |
| | 3-31-92 |

This way there is no question about test schedules, or excuses for being unprepared. It also gives me a target date to shoot for. Furthermore, it helps the students plan their time appropriately. They will be the first to notice any major conflicts in your plans (i.e., you've scheduled their short story test during spring holidays or on the same day that the band will be performing at the Special Olympics or on the same day that the students are having a huge physics exam) so that you can make accommodations accordingly.

- Have extra materials. In August when the back-to-school advertisements start materializing in the media, and school supplies are discounted, I purchase a supply of the basics: a gross of #2 pencils, six packages of notebook paper, and six packages of ballpoint pens. This usually covers nine months' worth of emergencies. Now it is true that Sam Walton (or whoever) gets about $10.00 of my hard-earned money. In the long run, however, it is well worth the investment to avoid the aggravation and wasted time caused by students who forget their supplies. Try to have an extra textbook or two on hand, also.

- Expect the unexpected. You thought the class would grasp the method of characterization in one class period. It took three days to get the information across. You thought it would take the class forty-five minutes to do exercise six. They finished in ten minutes. There were fifteen minutes left in fifth period and you were right on schedule, then there was a fire drill. As important as planning is, flexibility is of equal importance. Have several ten- to twenty-minute activities in mind to occupy unexpected time. Deal with interruptions rationally.

- Substitute plans. Sometimes you do know in advance that you are going to be absent, but frequently you don't. It is a good idea to have two or three days' worth of generic lesson plans prepared, complete with handouts, etc. Put the plans in a folder, clearly labeled. Put the folder in an accessible spot and alert the teacher next door and a reliable student or two as to its whereabouts. When Montezuma's revenge strikes at four in the morning, you can concentrate on survival without worrying about your classes.

- Paperwork. It never ends. In fact, it multiplies. There is a form, probably three, due in triplicate yesterday. I've spent better than two decades trying to devise a way to eliminate it with previous little success. The best you can do is try to minimize it. You'll find your own ways to deal with it.

   For starters, don't assign busywork. Homework and classwork should be meaningful practice. By the same token, it should be practice, not a threat. My method is something like this: when an assignment is made for homework or classwork, it is designed primarily to allow students to practice a skill which they have not mastered yet. I do not expect perfection. Consequently, if a student completes the entire assignment and turns it in on time, he receives a full credit. I look over the work to see that everything is complete and actually check two or three items per paper for accuracy. Each classwork/homework assignment is worth ten points (compared to a major test which is worth 100 points). Ten such assignments are made throughout the grading period, making a total of 100 points, the equivalent of one test grade. At the end of the marking period, I simply add up these points. I use the same technique with vocabulary work, first draft compositions, etc.

   Keep a copy of everything (in those handy looseleaf notebooks) appropriately labeled. While you probably won't use that exact same test or letter of recommendation again, you'll at least have a head start.

- Concentrate on the task at hand. While in the classroom, you are on stage. Put everything else out of your mind. Devote your attention to each task at hand. Don't allow your mind to wander to your empty propane tank or your son's science project. If you are out of focus, your students will be too.

- Solicit student input. Make a suggestion box and allow students to contribute ideas. True, most of the suggestions will be worthless, absurd, or obscene. Every once in a while, however, one of those inventive little creatures will propose an idea that really makes good sense for you. Capitalize on their perspective and perceptions.

*Anonymous, Grades 9 – 12, Georgia*


## CHAPTER HIGHLIGHTS

The teacher is the one who sets the learning pattern for the day, so you must be ready when the students arrive. To keep attention on the lesson, vary your presentation techniques. An interested student won't be a discipline challenge. Never walk into the classroom unprepared.

## CHAPTER 15

# *Assessing Student Work*

COMPARE teacher perceptions of the importance of assessing student work at the elementary level as opposed to the secondary level, and you'll note that elementary teachers rank it at the bottom of their list of top ten challenges, while secondary teachers rank it fifth. What accounts for the difference? It could be that at the elementary level, teachers feel it is more important to assist learners than to evaluate them. Secondary teachers may, at some level — perhaps subconsciously — be preparing their students for entry into a competitive society.

Whatever the reason for the difference, there are some assessment fundamentals that are important to remember for both elementary and secondary teachers.

First, the teacher needs to know just what it is he or she wants to assess. Make certain your students know your expectations of them, and how you are going to assess their performance in trying to reach your expectations. Don't be afraid to set high standards.

Now that the mechanics of your assessment package are in place, it is important that you communicate such to your students at the earliest possible moment. This is one of those things that you simply do not put off until later. In fact, the astute educator not only presents this package as soon as possible, he or she will provide an ongoing message regarding such. It's like doing a good commercial for a consumer product, you present your message frequently and it is less apt to be forgotten. The same holds true for your assessment program in your classroom.

Once you have established the format and communicated it to the students, you are obligated to stick with it. You do not have the luxury of changing the rules after you start the game.

Last but not least, you are also obligated to acknowledge the work your students provide. There is nothing more frustrating for a student than to submit work to the teacher on a regular basis and never receive any feedback on the work. There is no need for the teacher to have a life dedicated to feedback, but the rule is simple and pretty-well endorsed by teachers — if you assign it, then you need to assess it and return it. That seems fair.

Now give your undivided attention to our veteran exemplary teachers as they share their ideas on assessing student work.

---

Students should be told exactly how their work will be assessed. Tests, quizzes, classwork, homework and participation percentages should be explained on the first day of class. Project work may be jointly assessed by the students and the teacher. Whatever method of assessment is used, it must remain consistent. The students and parents need to know that the teacher has a policy which is available.

Nancy D. Ruby, *Secondary, Vermont*

---

As a high school mathematics teacher I have found that it is important to assess students' work frequently in order to immediately correct any misconceptions or misunderstandings. I do this by collecting homework (that the students have corrected) on a daily basis, quizzing students on an average of once a week, and then testing students at the end of a unit. I feel that it is also extremely important to return students' quizzes/tests to them the day after they have been completed in order to correct any problems before continuing on to the next unit.

Elaine W. Grant, *Grades 9 – 12, Maine*

---

Expect great things on a daily basis. All work/effort should be acknowledged and rewarded in some way. Emphasize learning from mistakes; give many credit/no credit opportunities before testing. Require a classmate's signature on work and hold both responsible for the accuracy. Often talk personally with students about their work. Grades/tests must be fair, accurate, logical, and defensible.

Elizabeth Bishop, *Grades 9 – 12, Michigan*

---

Set your standards for each grade and give these standards to the students. That way, you both know the guidelines. Occasionally allow the students to recommend standards. Assuming you're working on pacing and accuracy combined, ten math problems worked correctly in ten minutes or less is 100. Ten math problems worked correctly in fifteen minutes or less is 98, 97, etc.

Myra M. Gottsche, *Grades 11, 12, Mississippi*

You do not have to grade everything that a student does for your class. There is considerable merit in having them practice a skill — your grading it does not make it any more or less valuable. *Check and chuck* is a good motto for lots of things.

When an assignment is *really* important (a final draft of an essay, say, or a project notebook) grade it accordingly. Spend some time with assignments and don't let the kids rush you into turning it back the next day. Write comments on post-it notes to stick all over the work (this way the comments can be seen, but aren't a bloodbath). Don't be afraid to be critical *or* complimentary.

Sue Vaughn, *Grades 9 – 12, Nevada*

My attitudes about assessment have changed the most since I started teaching. I recommend reading Glasser's *Quality School* and taking classes or going to conferences on outcome-based education. Even if one does not believe totally in either philosophy, the tenets of each make one think. It's difficult to keep doing what one has always done in the face of these philosophies.

Now, before I teach anything, I decide what I want the students to learn. I am very specific about this. At the same time, I decide how I'll assess whether or not they've learned. Then, I tell the students what I am going to teach them and how I'll assess whether or not they've learned it. In addition, more and more, I also ask them to assess their learning. For example, after establishing what I mean by a quality paragraph, I'll do something so simple as asking them to place a Q on their paper before they hand it in, if *they* believe it's quality work. Both my standards and theirs have risen since I've started doing this.

Finally, I give them multiple opportunities to learn. Just as I no longer focus on covering vast amounts of material in too little time, I no longer focus on time itself.

Christine Robinson, *Grade 7, Minnesota*

Writing teachers can cut their grading load by at least half by having students keep writing portfolios of their work in the classroom. Students could be assigned a specific number of writings per week, and at the end of the week the teacher will have the student select the writing which he/she feels is best. The teacher can also grade the amount of work included in the portfolio. A quick observation will assure the teacher of the student's accomplishments.

Kay Neal, *Grades 9 – 12, Idaho*

I average each test, quizzes, and homework as the same percentage. For example, if there are three tests during the grading period, each test, homework and quizzes will count twenty percent. I use points for homework and convert to a percentage at the end of the grading period. I drop the lowest quiz grade during the grading period if a student takes all quizzes, or allow the student to miss one quiz per grading period with no penalty.

Gayle Blunier, *Grades 10 – 12, Indiana*

I make an assignment every day. It is always checked in class or collected and checked by me. I record grades every day. It is expected to be in on time. If it is one day late, it is worth half credit. If it is over one day late, it is a zero paper. Sometimes we have pop quizzes for a grade, or for extra credit. Tests are given after each chapter. They are the only grades I curve to the highest score. Extra credit points are available at times for doing work that is more difficult than regular classwork. Students' total points are added at the end of the quarter and a percentage is figured based on points possible. Classroom participation and a student's work ethic and attendance figure into their final grade by perhaps raising an 88.9% to a 90% for an example.

*Anonymous, Grades 9 – 12, Missouri*

Be consistent. Have high expectations and let them know it. Keep complete records. Keep old tests and other major grades in a folder for each student until the end of the school year. Let them know in the beginning that homework, or notebooks, or class participation will count as part of their grade.

Barbara B. Bryant, *Grades 8 – 12, Alabama*

A student should be assessed in three areas; verbal responses in class, classwork/homework, and tests. I find a weekly seating chart for each class with space for notes helps record verbal and visual responses during class.

I grade classwork/homework per day as one grade and only grade the work as the student tried or did not try. Tests are formally graded as right or wrong. Then the grade per grading period is one-third classwork/ homework, and tests are two-thirds of the grade. If the student needs extra points, look at classroom responses.

I find in most cases the student that is prepared in class does well when test time comes along, but keep in mind each response in class should be acknowledged.

Elaine Puckett, *Grade 8, Georgia*

At the beginning of the year I show students how I want their work done, how lab reports should be written, how daily assignments will be evaluated, and how tests will be graded. I also explain to them how each type of work is weighted and how final grades will be computed. I have periodic conferences with each student, as needed, concerning their work, and will share their grades with them at any time.

Howard L. Meserve, *Grades 11, 12, New Mexico*

I assess students' work on individual effort and achievement. A short individual meeting periodically helps the student keep his/her goals and achievements in mind.

Effort is the maximum weighted grade in my teaching areas. I've found emphasizing and rewarding effort enables even the poorest student to try harder.

Thomas D. Pennington, *Grades 6 – 12, Maine*

Because I teach Spanish by total immersion and allow only Spanish to be spoken in the classroom, it is easy to assess the students' competency. Each student is required to speak at least twice during each class. Many speak more often. I can easily see who needs extra help. Total involvement with each student is the key to good assessment of students' needs.

Paul C. Worthen, *Grades 9 – 12, Arizona*

Students need to be clearly told the criteria and procedure for grading their work. The criteria must be specific and complete. There must be no vagueness if class participation and behavior are part of the grade. Grading criteria must be made clear before, not after, students start an assignment or take a test. You should repeat the grading criteria frequently and if certain information is not stated, it should not be used. This would not be fair. Using a computer program to report grades on a weekly basis or after each major test works well, and this way students are never in the dark about grades.

Ron Adcock, *Grades 9 – 12, Iowa*

Assessing student work must be more than grading objective tests. Class discussion, oral reports, short answer and essay tests, and papers tell much more. Teach and look for examples of the higher levels of thinking – critical analysis and creative thinking.

David Nienkamp, *Grades 7, 10, 12, Nebraska*

The student should have every opportunity to illustrate his or her learning. They should not be assessed only on test grades. The students should be able to do both oral and written reports, do projects either individually or cooperatively, or even make posters to illustrate their learning.

If we must grade, and I believe we must, it should be as fair as possible. Every adjustment must be made to ensure fairness. Students with learning disabilities and students with reading and other problems must have adjustments made to ensure that they are tested and evaluated fairly.

Charles J. Daileanes, *Grade 8, New Hampshire*

---

A child should know what is expected of him/her on classwork, homework, and test papers. Give point values for sections, and state clearly what you want. For compositions, I use a focus correction area. All students' papers carry the focus at the top. In it I state those areas I intend to focus on when I am correcting. This focus serves two purposes: first, it allows the poor writer a chance to concentrate on specific areas; second, it helps me assess writing with specific purpose. As the student progresses, the focus gets longer and more involved. Instead of a blind subjective writing assignment, the student can hone in on the skills required in the focus.

Beverly-Ann Gill, *Grade 9, Rhode Island*

---

Nowhere is it written that teachers must assign a letter grade to every assignment. But if it is important to assign, then it is important that the instructor at least provide some type of positive feedback to the students. Also, assessment doesn't always mean a paper/pencil test. Students can certainly demonstrate mastery in other ways. In addition to the traditional test, offer other activities, such as skits, artwork, debates, creative writing, etc., which will help in assessing the students' understanding of the material.

Terry Boyle, *Grades 10, 11, Virginia*

---

In the field of publications, I must be very careful when I assess student work. Since each child works individually, I have to grade individual assignments. Grading becomes less subjective with a point system. The editors of the publication also help monitor other students, so that editors, photographers, salespeople, and artists are judged accordingly. Each student is given a fair chance to excel.

Marilynn Sexton, *Grades 9 – 12, Montana*

Because student evaluation can be a touchy subject with parents and students, it is important to employ a variety of techniques to meet every type of learner. This ensures that every student has the opportunity to succeed in some manner. Use oral (speech, presentation) and written (essay, true/false, multiple choice, short answer) tests to assess student progress. Give full credit for simply completing a project when applicable. Use focus correction areas or primary trait scoring when appropriate. To increase healthy tension within the class, have competitive scoring on occasion. Variety gives balance to your assessment.

Kimberly Sturgeon, *Grades 9 – 12, Virginia*

## CHAPTER HIGHLIGHTS

There are many grading techniques that can be used. Whatever technique you choose, the average grade of the class just might be reflecting upon your ability to make the material understood. Take other factors into consideration, such as classroom participation. Be fair and let your students know how you'll be grading.

## CHAPTER 16

# Relations with Colleagues

YOU know the value of having a good working relationship with the people around you. For the first-year teacher, developing good relationships with fellow teachers and administrators is a special challenge. One helpful approach is to take the initiative. Don't wait for your fellow teachers to come to you, go to them. You must do your part by establishing your commitment to the team early.

As a teacher, you have three main resources—your community, your school's physical assets and your personal relationships. If your colleagues aren't in your resource base, you will probably never become a whole teacher. Don't lose sight of the fact that your school administrators need to be included within your circle of colleagues. Be wary of promoting a schism between teachers and administrators.

Recognize the fact that in your teaching career you may find one or two people with whom you just cannot get along. That's okay, so long as you honor them as teachers and respect them as individuals. Or as one teacher said, "It's alright to disagree, but it's not alright to be disagreeable."

Our relationships with our colleagues are reflected in our relationships with students. If we do well with our colleagues, we'll do well with our students. Another point to remember is that students seem to have a sixth sense about relationships. When things aren't going well, they'll sense it and it may interfere with your classroom work.

---

Be a teamworker as soon as you walk through the doors of the school. Lend a helping hand whenever you see an opportunity. This will help you establish good relations with the other faculty, as well as orient you to the workings inside the school. Seek to be good at what you do and look for other examples to observe. A lot of practical knowledge can be gained from good role models. Approach your peers with sincerity. Recognize the extra things you see them doing. Counsel with those you trust, and seek to build a supportive professional relationship with them.

Kimberly Sturgeon, *Grades 9–12, Virginia*

In a school environment where people mutually respect each other, a positive, caring climate is established which directly affects the students. It is helpful to understand that we are all a community of learners whose goals are the same, but whose paths to those goals are as varied and unique as the children being taught. Honor the diversity within your particular school and you'll be richer for it.

Terry Boyle, *Grades 10, 11, Virginia*

Never talk negatively about colleagues to students, other colleagues, or parents. Never expect or demand special privileges that are not available to all colleagues. You must respect the curriculum in all areas. The attitude that you care about science and couldn't care less about physical education negates professional courtesy. You must support leadership at all levels within the school. Never disregard or degrade the authority of anyone in a leadership position. Respect all the people who work in the school. At least extend professional respect, if you can't like them.

Ron Adcock, *Grades 9 – 12, Iowa*

I show support for my colleagues by attending their activities (concerts, athletic activities, etc.). I interact with them outside of the classroom when possible. I always try to offer an encouraging word. I work closely with many of them in extracurricular activities (student council, coaching, and forums). I am always willing to assist with their activities. They, in turn, support me in mine. I try to show an interest in their fields. I sincerely care about them and wish them success. Honesty and sincerity is essential when working with others.

Paul C. Worthen, *Grades 9 – 12, Arizona*

No two people completely agree; certainly not teachers. Though one may not be in agreement with a colleague, one's fellow teacher is part of the total team. A full appreciation of what other educators teach is essential. Attending each other's events, meets, concerts, matches, and games can help all departments stay amiably on the same team.

There is no "most important" subject in a curriculum. Equal respect for all subjects and the people who teach them is vital. No teacher can educate alone. Learn to share the good times and bad with fellow colleagues.

Thomas D. Pennington, *Grades 6 – 12, Maine*

I feel that each of my colleagues, whether certificated teachers or noncertificated staff, should be treated with the same respect that I would like. They all should be treated in a professional, warm manner as ladies and gentlemen. Help should be offered to those beginning their careers. Students are well aware of the relationships among staff members, and each of us should strive to act as positive role models for the students.

Howard L. Meserve, *Grades 11, 12, New Mexico*

---

The relationship between colleagues is a give-and-take situation. Remember, everyone is willing to help a new teacher if it does not become just a taking situation. Every teacher has something to offer. Share what worked when you were a student teacher, or what you may have learned during your classroom experience in college.

Also, always remember if you are a member of a team do your share of the work. No one likes to do their work and yours also.

Elaine Puckett, *Grade 8, Georgia*

---

Be friendly, but be careful when asking for advice. A mentor teacher plan can be very effective. Asking advice from everyone will only confuse a new teacher. Never gossip about students or other colleagues. Words can return to haunt you. Learn to be self-sufficient and assertive; your problems will become minimal. Just because one teacher has problems with a certain student does not mean another teacher will have problems. It is each teacher's responsibility to help his/her students learn.

Kay Neal, *Grades 9 – 12, Idaho*

---

Do everything you can to avoid getting caught up in office politics. As a beginning teacher, you don't know (and probably don't want to know) the background and history of much of what happens at your school. You don't have to be a "yes" person to avoid making enemies—just be quiet or nonjudgmental until you're sure you have all the facts. Staying out of office politics will also save you *tons* of time.

Sue Vaughn, *Grades 9 – 12, Nevada*

---

Colleagues are among the best resources a teacher can find. Their experience, both past and present, can be of tremendous help. Most are willing to share methods, materials, and experience. They should be treated with respect. They are, after all, teachers who are, for the most part, always ready to teach, even their peers. Administrators should also be viewed as mentors. Most welcome the opportunity to share their experience.

Nancy D. Ruby, *Secondary, Vermont*

Accept the fact that you *do* work with them so you may as well learn to work *with* them. (Odds aren't real good that they'll be checking out tomorrow!) Accept the fact that, like all of life, some people will continually empty your tank while others constantly fill it. Recognize the worth and ability of each. Praise where deserved, and accept compliments when earned.

Jennifer Ortman, *Grades 9 – 12, South Dakota*

---

I feel many schools neglect this important area. Every inexperienced teacher or teacher new to a school should be greeted by their colleagues and made to feel welcome. I feel teachers are easily threatened and often feel insecure. With national attention directed to low test scores, it is understandable that teachers become defensive. To lessen defensive behaviors, teachers must learn to communicate frequently, and to share experiences. Classrooms should be open, and colleagues invited in frequently. Schools should implement a program where exemplary teachers guide others toward positive changes in classroom methods and discipline. If no such program is in place, a new teacher should seek help from colleagues on his or her own.

Kyle Ann Morrow, *Grades 7 – 12, Alaska*

---

Seek the people in your school who love teaching children, and catch their enthusiasm. Take your difficult problems to them and ask how they resolve similar situations. They will almost always be flattered that you asked, and generous with ideas and insight. Find someone who will be your professional partner in talking about classroom practice and curriculum. If your partner will engage in peer coaching or keep a dialogue journal with you, so much the better. Most importantly, find the teachers who understand that the heart of good teaching is career-long learning, and join them in that process.

Lynne Cullinane, *Grades 9 – 12, New Jersey*

---

Not to sound negative, but the best advice that I can give to a beginning teacher is, "Teachers tend to be the most unprofessional of professionals."

Frank R. Gregory, *Grades 9, 12, Delaware*

---

Don't be a chronic complainer, and be willing to do your share of work – whether it's cleaning up in the lounge, counting books, chaperoning a dance, or whatever. Remember to smile!

Sally Tanner, *Grades 10 – 12, North Dakota*

---

To begin with, I am a friend. I am willing to share my materials with the new teacher or with the veteran teacher who is teaching something difficult. I have tried to help even those outside my discipline with problems they have had in organizing their work or with discipline problems as they have asked me. I have never been one to shirk responsibility or to point fingers at those who do. Race has never been a problem, either, for I respect every person and am willing to work with anyone. I also do not like cliques among teachers, because it alienates those not in the group.

Nell W. Meriwether, *Grade 12, Louisiana*

---

Every school setting can be compared to a home filled with a large family. Teachers and other staff are the "parents" of the many children and young people. In order to maintain a healthy home for growth and learning, the teachers must be healthy, and this means that teamwork is essential. Furthermore, maintaining positive collegial relationships not only helps students function well, but will also help you feel good about your work environment. Human nature will attract you to some colleagues rather than others. It is in these relationships that you will find support and encouragement when you have a conflict to solve, when you're feeling down about a particular student interaction, or a class that didn't work out as planned. Positive relationships with colleagues are important and necessary.

Brenda M. Naze, *Grades 9 – 12, Wisconsin*

---

It is always important to remember that your teaching colleagues are professionals. They expect to be referred to in that manner and will expect the same behavior of you. You should respect their opinions on teaching issues, and the manner in which they conduct themselves in the classroom.

If a problem should arise between you and a colleague, it is wise to discuss the situation yourselves first. Try to reach an understanding that you can both respect.

Kelly Schlaak, *Grades 9 – 12, Minnesota*

---

Most noticeable to me as a new teacher was the lack of interface and support I had anticipated to come from my peers. There was not nearly the organized opportunity for good one-on-one note comparing I thought there would be. As it likely is in most professions, most teachers' lounge and lunchroom discussions are not positively oriented, but tend to be complaint sessions. And those planned ''new guy'' in-services never seem to be designed to work on any of your problems. But look around. I got to school an hour and a half early one morning and bumped into a klatch of early risers who habitually gather each morning in one of the science rooms to drink coffee and talk business. I made myself a member of that informal group, started bringing donuts on my designated Friday, and learned a tremendous amount about teaching, the system, and about my students and their families through the eyes of other teachers. Be careful; you may well be allowed to work in a vacuum. It will be up to you to see that doesn't happen.

Ken Clark, *Grades 11, 12, Ohio*

---

Compliment your colleagues often but only for worthy accomplishments. Share your ideas and materials with them. Ask them their opinions and ideas. Do not get into gripe sessions. If you borrow materials, return them. Ask before taking anything from their rooms. Show enthusiasm for your profession. Put in a full day's work. Keep your room and desk clean. Confront a colleague directly and assertively if problems arise. Always have a good sense of humor.

Thomas Koenigsberger, *Grades 9 – 12, Illinois*

## CHAPTER HIGHLIGHTS

Give your support both in and out of school. Be ready to listen and help out, and always say positive things – or say nothing at all – about your colleagues.

## CHAPTER 17

# Dealing with Individual Differences

JUST as you will notice differences between colleagues, and between yourself and colleagues, so, too, will you notice differences between students. To reach your potential as a teacher, and to help your students learn to the best of their abilities, you've got to use "different strokes for different folks" in the classroom. It helps if you bear in mind that you teach for the student's sake, not for the teacher's. And it follows that if you teach differently to different students, then you must assess them differently.

Hard work? Yes. Worth it? Absolutely. Here, from the trenches, are some helpful tips.

---

Provide multiple opportunities for success, and multiple ways of attaining it. Some children excel at individual pencil-and-paper tasks: tests, essays, reports, and worksheets are great for them. Some need to talk about what they are learning: cooperative learning groups, small- and large-group discussions, oral reports, dramatizations and role playing can all help. Some need to work with their hands on concrete materials: models, diagrams and charts, laboratory work of all kinds appeal to these students. In addition to working in areas where they excel, students offered a wide range of learning activities are also being offered a wide range of opportunities to stretch and grow.

Lynne Cullinane, *Grades 9 – 12, New Jersey*

---

In dealing with differences and problems in individual students, a teacher should remember that many students are affected, or were affected at one time, by drug and/or alcohol abuse. Whether the student experienced these effects as a fetus if the mother drank or did drugs, through family member use and abuse, through association with imbibing friends, or through their own personal use or abuse, a teacher must remember that such students respond negatively to loud, aggressive or high pressure situations.

A positive response results when the teacher remains calm and shows understanding toward the student. Also, a teacher must remember that such students don't intentionally forget class materials or homework. They do want to be successful and fit in with their peers. They just need repeated simple directives and frequent reminders. Just the fact that they are in school may be considered a positive move for many of these kids.

Jennifer Ortman, *Grades 9 – 12, South Dakota*

---

We must operate on the premise that all individuals learn differently; therefore, we must vary our instruction to meet the needs of all learners. Balance lecture with cooperative learning strategies so that instruction will be teacher-centered part of the time, and student-centered the other part. Likewise, offer a variety of activities to measure the mastery of content so that all learners can demonstrate what they know.

Terry Boyle, *Grades 10, 11, Virginia*

---

The individual student is going to learn at a rate that will fit his or her learning style. What the teacher needs to do is keep the interest of the advanced student, without losing the students that are struggling. One way to achieve this is to use the resource of these students. Have them tutor struggling students. You can also use the various cooperative learning activities which allow a group to move together.

Dan A. Fenner, *Grades 11, 12, Pennsylvania*

---

As a math teacher, I encounter a wide range of individual aptitude and interest in the subject. I try to make use of the talented and motivated students to lead the way in discovery-method instruction, and to explain as much of the subject to other students. At the same time, one must guard against the tendency to overemphasize the advanced student and neglect the average student.

David Hungerford, *Grades 9 – 12, New Jersey*

---

My business math class is a good example when one must deal with differences of students. Some students have difficulty with fractions, and for others calculus is not hard. I try to teach somewhere in the middle. For slower students, I give extra time and help. For faster students, they can work ahead. I try to relate where the student is. For example, if one likes sports, then I'll relate things in a sports way.

Bob Eschmann, *Grades 9 – 12, Illinois*

Plan lessons, activities, and teaching strategies so every child will be able to feel a sense of accomplishment. Be mindful that an excellent rating for a child with limited ability might be merely satisfactory for an advanced student.

Elizabeth Burton, *Grade 9, Tennessee*

I prepare varied-level assignments. All students receive the same assignment sheet. However, not all students are expected to finish the assignment. The sheets are prepared with the easier questions at the beginning and building to greater levels of mastery. Students of higher ability will finish and receive an A if successfully completed. Students with less ability may only successfully complete half the assignment and also receive an A.

Pamela A. Reed, *Grades 9 – 12, Michigan*

I prefer mainstreaming students, but the resulting classroom mix is a challenge. Team teaching with a special education teacher has helped me the most. Often, what we teach is what has always come easily to us. Therefore, we need to educate ourselves as to the variety of roadblocks students bring to learning our subject.

I've learned to ask questions of the special education people. I've learned to keep an open mind and be receptive to any idea that works. I use my knowledge and experience with the possibly more dramatic roadblocks for the special education students to remind me that each one of my students has special needs; they're all different.

Christine Robinson, *Grade 7, Minnesota*

I try to use those students who are able to complete an assignment with a minimum of direction from me as student helpers. I will have these students help another student who is having difficulty completing an assignment. I try to assign enrichment problems for those who always complete their assignment before the other students not just more of the same type of problems but challenging problems and for which they will be rewarded both intrinsically and extrinsically (extra credit points).

Gayle Blunier, *Grades 10 – 12, Indiana*

I have found that a few minutes before or after school with a student who is having difficulty really helps. Nothing takes the place of a little individualized attention. Allowing students to communicate with each other during class helps too. For example, you could tell your class, ''For the next fifteen minutes pair up, compare answers, work problems through together that you disagree on, to correct them.'' I might pair up a person having trouble with a person who has their work all completed and knows what they are doing (peer coaching). Sometimes I send students to the board and talk them through a problem. Sometimes I send students to the board as partners to work through a problem together.

*Anonymous, Grades 9 – 12, Missouri*

---

Working in groups to permit peer tutoring will help compensate for individual differences. Those who tutor will reinforce their knowledge and expertise. Those who need tutoring will receive individual help which the teacher may not be able to provide. Groups should be heterogeneous; a mixture of ability levels, different races, different sexes. Change groups every grading period. Make them responsible for working cooperatively and helping the member of the groups. You may keep a group grade average and give rewards for improvement.

Barbara B. Bryant, *Grades 8 – 12, Alabama*

---

The solution to this is easily stated but hard to achieve. Make it challenging and difficult to achieve the A; make the C achievable for any student in the class; make sure special students are involved and do learn from the class material. Know how hard to push and how hard not to push for each individual. Always demand effort and keep the standards high.

David Nienkamp, *Grades 7, 10, 12, Nebraska*

---

Our job as teachers is to find what works for each student. We must find ways for our students to succeed rather than a reason to fail them. This could be achieved by having a variety of different assignments. Grading should not be done purely from testing. Poster projects and presentations are other useful tools to help students with individual differences to be successful.

I have used learning partnerships, peer tutoring, and cooperative learning assignments to meet all students' needs. I have also had students work with specialists in and out of the classroom in an effort to improve themselves.

Many different students can function successfully in the classroom under the right circumstances, and with the necessary support network.

Charles J. Daileanes, *Grade 8, New Hampshire*

## CHAPTER HIGHLIGHTS

All students learn differently. The task of the teacher is to recognize these differences and then capitalize on them so each student can be successful. You may have to challenge students on different levels to be sure everyone is working to his or her potential.

## CHAPTER 18

# Dealing with Problems of
# Individual Students

BEFORE you can deal with your students' problems, you need good baseline data on each student. Don't confuse this with developing a secret file on everybody. The first step is get to know the students' names—fast! Next, get acquainted—schedule a short visit with each student just to talk. Then, be alert to changes in a student's behavior.

When you spot something that you believe is worth pursuing—make your move! Get with the individual who you suspect has a problem. *Listen* to what he or she might have to say. Remember, this individual may or may not speak to the issue. If you learn something, good. If you hit a wall, don't despair. Stay alert and be sure to make your availability known.

Finally, assess the situation and then decide the extent to which you should be involved. At the very least, you have three options. First, you may close the issue after you and the student have collaborated on it. Second, you may find it in your best interest, as well as the student's, to have the parents of the student participate in the matter at hand. Third, you may choose to pull the guidance counselor in. Each situation must be dealt with on an individual basis and then let your intellect and conscience be your guide.

Here are some ways our veterans handled a wide range of personal situations.

---

Notice changes in attitude, behavior, or performance of a child. Ask the child in a one-on-one setting if there is anything he would like to discuss. Listen attentively. Do not judge, criticize, or embarrass the child. If the situation warrants, refer the child to the guidance counselor, who can provide additional help, if needed.

Brenda Clemons, *Grades 10 – 12, Virginia*

---

Administer a personal inventory at the beginning of the year, and get to know your students individually. Confer with them all on an individual basis periodically, regardless of their standing in the class. Let them know that you care about their academic as well as their personal problems. Allow them to feel comfortable to discuss their problems with you. Poor performance and erratic behavior signify problems, and should be dealt with immediately.

Abbas Saberali, *Grades 7 – 9, Texas*

---

Not all children learn at the same pace. However, not all children will feel comfortable enough to ask you for help. It is sometimes easier for them to fail a test than to ask for help. Therefore, after checking to see that everyone has the work done, I divide the class into various groups with at least one or more gifted students on the subject included within the group. Individuals within each group check each other's papers. If a question arises, the group consults me. That way, no one is singled out. I also conduct an ''English Night'' after school on Wednesdays, where students may make up work missed, come for help, or just come in to talk. You'd be surprised how many conversations have nothing to do with English.

Beverly-Ann Gill, *Grade 9, Rhode Island*

---

Every student is an individual. Every student has strengths and weaknesses. A teacher has to try to understand each student and discover his skills. I try to learn names immediately, and then I try to visit with each student. I want the teens to know that I care. I encourage all of the students to be comfortable in the classroom. I want them to be active participants in learning. I try to notice all of them by calling on each of them each day. Teens develop self-esteem when you are aware of their interests. I often ask questions and offer encouragement. Frequently I ask students to assess my teaching skills. I ask them to critique me and tell me how they could have learned easier. At times, I need to change my teaching styles so that they complement my students' learning styles.

Marilynn Sexton, *Grades 9 – 12, Montana*

---

When you have identified a behavioral or academic problem with a student and can give evidence of it, engage in a one-on-one discussion with that student. Decide mutually on a plan of action to solve this problem. Inform the parents – and administration if necessary – of this meeting. Have follow-up discussions periodically. Note improvements so the student can be encouraged. Continue to point out the rewards the student will reap if this problem is worked out. Also, these one-on-one times will meet any problem that is a ploy for attention.

Kimberly Sturgeon, *Grades 9 – 12, Virginia*

---

Finding personal time (or having the expertise) to deal with problems of individual students is difficult at best. Along with talking to, or leaving notes for support staff about students with whom I have concerns, I also try to check in with the students myself. For me, this happens before or after class, or perhaps in the hallway or cafeteria. Even a simple, "How's your day going?" "How are you feeling?" or "Do you need some extra help?" makes a difference. In classes where students keep journals, I might write a note to those for whom I have concerns. Your interest and concern will let students know you care. This alone can help them stay more focused on learning when working with you, and in the school setting as a whole. Giving students a chance to release some of their feelings about issues in their lives, even if only for a few minutes, has an impact. Be aware that by not recognizing or addressing problems of individual students, these problems will probably escalate rather than disappear.

Brenda M. Naze, *Grades 9 – 12, Wisconsin*

---

Each school day a teacher is in contact with many students. They all have different personalities and come from a variety of backgrounds. It is important to remember that they are human and can have good and bad days. Try not to put a label on a student in reference to their academic capabilities, peer group, family background, etc. The majority of students want to be in school. They are excited to learn and make plans for a future.

When a case arises where the teacher needs to work with a student who is, or has been, a problem, the teacher must remember to listen to the student first and then, must consider all options for dealing with the situation. Try to reach a mutual agreement on how to correct the problem.

Kelly Schlaak, *Grades 9 – 12, Minnesota*

---

When a problem arises (for example, a fight, or a student who gets an F on a test), I try to find out background information. I want to know what is causing this to happen. That means one has to be generally concerned and also be a good listener. One should interact in a respectful manner with students. It sure helps solve problems if they know you are a friend in addition to a teacher.

Bob Eschmann, *Grades 9 – 12, Illinois*

A sort of triage takes place—there are those students whose problems are not too bad, those whose problems are bad but treatable, and those whose problems are totally overwhelming. As in war, it is the second category that gets the most attention; parent contact, individual/student talk sessions, etc.

David Hungerford, *Grades 9 – 12, New Jersey*

---

I teach high school. As a result, I find I have a much better time getting information across to the students if I treat them with respect. No one likes to be talked down to. Nothing will cause separation between you and your students more than if they perceive you as only a superior.

Dan A. Fenner, *Secondary, Pennsylvania*

---

Listen with all five senses! Because my foreign language students are asked to describe or talk about many situations, several events have unfolded. I make it a point to follow up on various comments. For example, a student always uses negative adjectives to describe her father. Another says she isn't sleeping much at night; and another is living on Tums for an upset stomach. I find one-to-one time with that student and get around to asking if they want to share or explain these comments. About ninety-nine percent of the time, the student is harboring a deep family problem or difficulty in a personal relationship. Don't ignore repeated negative or abnormal comments. Listen and follow up!

Jennifer Ortman, *Grades 9 – 12, South Dakota*

## CHAPTER HIGHLIGHTS

When problems arise, either academic or personal, take time to listen on a one-on-one basis. Everyone needs a friend who can help directly or get one going in the right direction to get help.

## CHAPTER 19

# Heavy Teaching Load with Insufficient Preparation Time

YOU simply will not have enough hours in the day to do all that needs to get done. However, if you establish good relations with your colleagues, you will have help available. Don't expect your colleagues to do your work for you, but what you can expect is for them to offer you a variety of insights and maybe a few shortcuts.

You can also get help from students and parents. They are willing to do some of your work. They can do things like grade papers, provide classroom labor, and even do some of your take-home projects.

Regardless of how much help you can dig up, you are still going to find yourself short on time. The most important thing to remember is that you are not alone, and whatever you do, don't let the situation cause you to self-destruct. As the exemplary teachers suggest, a heavy teaching load (especially common for first-year teachers) with insufficient preparation time comes with the territory.

---

Take advantage of every moment. Prepare lesson plans as far in advance as possible. Give assignments where the students can exchange papers and check them. Use checklists for evaluating student's work. Grade papers and record grades immediately. Make sure that all assignments do not generate lots of papers to be graded at the same time. The secret is planning effectively and grading immediately.

Brenda Clemons, *Grades 10 – 12, Virginia*

---

For four years, I had four different preparations a day. I taught English, introduction to journalism, yearbook class, and newspaper class. I had to earn money to pay for the newspaper and I prayed that the yearbook would pay for itself. My students also produced school newsletters, literary magazines, and athletic programs. We frequently designed activity tickets and season athletic tickets. There was never a dull moment. Some of my fellow teachers questioned my sanity. They wondered why I wasn't more organized. They asked why I spent so much time at school. An associate principal asked me if I had a life.

I actually enjoyed what I was doing. I had a new computer lab and I was learning something new every day. Each day was an adventure. I did act like a space cadet when I was so busy but it was a great experience. I felt that the students were learning some valuable skills.

Marilynn Sexton, Grades 9 – 12, Montana

---

It's important to remember that this is both a political and a personal issue. That's one of the reasons it is so frustrating. Politically, it means getting involved in the management of one's school. Tell the people who set the schedules what matters to you; be diplomatic but direct. You still may not get what you need, but you'll feel less powerless.

Personally, set realistic goals for what you can accomplish given your teaching assignment. Be creative and draw on outside resources; ask for parent helpers. Don't underestimate what your students can do; they actually feel flattered when you entrust them with bulletin boards, errands, filing papers—even cleaning boards.

Finally, relax and take time for yourself when you're away from school. You need to recharge, for your own sake as well as for your students'.

Christine Robinson, Grade 7, Minnesota

---

Persevere the first year. Get organized and keep every worksheet and test in an appropriately labeled folder. I have used the same grammar worksheets for grades nine through twelve. Many tests and worksheets can be used over and over, or they can facilitate your making out new ones. Considering that beginning teachers often get the "leftovers," you may be overwhelmed that first year. In addition to getting organized, you should not be afraid to ask other teachers for help in sharing materials and ideas.

Barbara B. Bryant, Grades 8 – 12, Alabama

---

Here's how I handle the problem of a heavy teaching load with insufficient time. First, I keep a notebook of everything I do for each class, writing down my lesson plans, activities, tests, etc. for each unit. That way, I have good plans that I only have to add to as I teach. Next, I give some assignments which require a minimum of grading but which are effective teaching tools. I also use the process method of teaching writing which involves peer evaluations and which cut down on some errors that might reach me.

Nell W. Meriwether, Grade 12, Louisiana

Keep in mind that every day is not going to be perfect. Be flexible and let the students show their strengths and weaknesses, then expand on your lesson for them. Stay with the set curriculum, but only use it as an outline. Always remember that what works in one class may not work in the next class.

Elaine Puckett, *Grade 8, Georgia*

I firmly believe that there is no substitute for a well-grounded knowledge of the subject matter being taught. If the teacher has the proper subject matter and knowledge, and sets aside time each day, then several different approaches to the lesson can be outlined, then adapted to the class and to the individual student. Time is valuable, and must be spent as carefully as money.

Howard L. Meserve, *Grades 11, 12, New Mexico*

When faced with a heavy teaching load and little time for preparation, I have found that I must squeeze every minute from every hour and work hard. No one who enters the teaching profession should expect to work only from 8 A.M. to 3 P.M. New teachers, especially, should be prepared to spend many hours after school and at home preparing for the next day. In order for a teacher to be effective, I feel that they must be well prepared, not winging it, even if that means late nights and/or early mornings.

Elaine W. Grant, *Grades 9 – 12, Maine*

See if there is a master teacher or someone you feel could and would be willing to offer guidance. Set realistic goals and know you cannot have it all well-organized and creative all the time!

Susan LaPeer, *Grades 9 – 12, Massachusetts*

The burden of preparing for effective teaching appears overwhelming to the beginning teacher. So many things to do and so little time to accomplish them. One secret is budgeting and planning time to plan. I carry a pocket calendar at all times. When things come up like faculty meetings, departmental reports, grades, unit plans, etc., I place these on my calendar. Each morning, I check the calendar and then prepare a three column list marked urgent, important, and future consideration. I fill in the column with the activities prioritized into urgent (these must be dealt with today – lessons for tomorrow, etc.); important (these need to be worked on soon – contact outside resource person for next unit); and future (things that need to be done, but not right away). Each day I try to do those items in the urgent column. If time allows, I'll do several in the important column also. The next day everything shifts to the left. This way, I have some control over the things that need to be done. Of course, not all urgent activities will be accomplished in one day. They stay in the appropriate column until completed. The same goes for the other two columns. Different people are more productive at different times of the day. Some people can complete tomorrow's work right after school while others may need a break before tackling tomorrow's load. Decide the time at which you work best – intense and uninterrupted – then set this as part of your daily routine. Make planning time a habit in your schedule.

Robert H. Yunker, *Grades 10 – 12, Maryland*

## CHAPTER HIGHLIGHTS

Nothing substitutes for a well-prepared lesson plan made out as far in advance as possible. Learn to use time wisely, and expect to put in overtime.

## CHAPTER 20

# Insufficient Materials and Supplies

ELEMENTARY teachers are plagued more by insufficient supplies and materials than are secondary teachers. Partly this is because hands-on learning is more pronounced in the elementary grades. However, hands-on learning is appropriate at the secondary level, as well. For all grades, there is help.

First, there is a myriad of *free* materials available, but you have to look for them and you have to work to get them.

Another approach is to improvise with what you already have. Go outside the lines.

Use your middlemen. They include students, the PTO, boards, administration, professional organizations, and advisory councils. And don't forget your community. It is loaded with materials and supplies that can be brought to you, or that you can go to.

---

Lack of textbooks and supplemental materials has been a problem for all teachers, old and new alike. One way to increase your resources is to start your own vertical file. I use my students as a primary source. They bring in the old magazines from home – ones that are to be thrown away. Once a month, I have all my classes scan for articles that pertain to the subject area. Sometimes, I create a specific list of topics. The students then cut out the articles and I catalog them according to the appropriate categories. A second benefit of this activity is that my students become aware of excellent reading materials. I also send a form to various people within the community asking if they would be willing to volunteer to be a resource person willing to discuss their area of expertise with my classes. Videotape the presentation to show to other classes during the day. This creates a resource call list. The list includes not only doctors and lawyers, but also truck drivers, housewives, coal miners, and retired people. They become a primary source of what is going on in the real world.

Robert H. Yunker, *Grades 10 – 12, Maryland*

155

Every teacher must be a bit of a scavenger. Never throw away paper unless both sides have been used. In math, the back of an old test could be geometric shapes in its next life. Old newspapers can be used in all subjects, besides keeping the student aware of the world around him. Never limit yourself to just the textbook. Do not be afraid to experiment with newspapers, guest speakers, paper of all kinds, and fast-food containers.

Elaine Puckett, *Grade 8, Georgia*

Observing good educators over the past twenty years, I've found that good teachers teach regardless of supplies and materials. When lacking money to buy what I feel are essential materials, I usually turn to my class for suggestions. Shall we have a fund-raiser? do without? share? improvise? Inter-departmental sharing can often save mass expenditures, e.g., audiovisual equipment, supplies, etc. I have bartered for supplies during hard times. Once my band marched for the grand opening of a discount store in exchange for some basic supplies our department needed badly. Be innovative. One way or another, good teachers teach!

Thomas D. Pennington, *Grades 6 – 12, Maine*

Do everything you can to get your school to pay for things, and borrow, borrow, borrow. I must say that I spend a great deal of my own money on materials for classes, but if I were in the business world, I daresay that I'd spend money on extras, too. Most importantly, *ask* before you buy. Administrators have a way of finding money if you have a legitimate need. Don't ask the day before you need it, go into the situation with all the facts and figures you need. Booster organizations are a wonderful source of funds.

Sue Vaughn, *Grades 9 – 12, Nevada*

Be creative, look to newspapers and magazines for ideas and materials. Contact local businesses and state agencies for materials such as maps, charts, field trips, etc. Don't be afraid to allow the students to come up with ideas and materials via their parents.

Frank R. Gregory, *Grades 9, 12, Delaware*

Report to your school building *before* school starts. Allow enough time for more materials to be ordered if you feel the supply is insufficient.

Waiting until the day school starts to notice these things puts you at a real disadvantage.

Sally Tanner, *Grades 10 – 12, North Dakota*

---

The primary means of getting around this problem is to utilize other texts, and similar types of resources. This takes time, but if you search long enough, you can find lessons and lab-related activities that will fit with your supplies.

Dan A. Fenner, *Grades 11, 12, Pennsylvania*

## CHAPTER HIGHLIGHTS

This is a problem for all teachers, but especially for first-year teachers because they haven't had the opportunity to build resources. Be creative, borrow, and save everything.

## CHAPTER 21

# Wrap-Up

TEACHING is exciting and rewarding. But it is also challenging, frustrating, and sometimes brutal. Teaching is a complex activity. It calls for a wide variety of cognitive and affective skills, many of which need to be implemented at the same time.

This book has provided you with widespread insights into some of the biggest problems first-year and entry-level teachers face, as provided by selected veteran, exemplary teachers. By no means are the problems focused on in this book the only problems a teacher will face, but if you can handle these big ones, you should be able to survive and even control the other challenges that come to visit your classroom.

You must keep in mind that there is no one remedy for every problem. Every teacher must identify which process, technique, or strategy works best for him or her. But what works today may not work tomorrow.

What you will find most helpful is a resource base of techniques and strategies to draw upon to make your teaching the best it can be. Developing this resource base is a career-long task. *Tips from the Trenches* was written to help you begin building that resource base.

# Elementary Teacher Directory

**Abell-Victory, Julie** (K − 12)
Stroudsburg Area School District
123 Linden Street
Stroudsburg, PA 18360

**Andrews, William** (K − 12)
Boylston Elementary School
200 Sewall Street
Boylston, MA 01505

**Baker, Barbara** (K − 6)
Hanalei Elementary
P.O. Box 802
Hanalei, HI 96714

**Baird, Sharron** (2, 4)
Oakland Elementary School
P.O. Box 390
Oakland, OR 97462

**Baysinger, Ron** (6)
North Elementary
63 Fletchall Avenue
Poseyville, IN 47633

**Bell, Barbara J.** (4)
East Pike Elementary School
501 E. Pike
Indiana, PA 15701

**Bell, Velma** (L.D., 3 − 5)
Gordon Elementary
1205 Metropolitan S.E.
Atlanta, GA 30316

**Berliner, Maribeth** (3)
Richmond Elementary School
R.R. 1, Box 551, Jericho Road
Richmond, VT 05477

**Bly, Frankie C.** (5)
Blue Earth Schools
East 6th Street
Blue Earth, MN 56013

**Bruton, Eileen** (5)
Salem Consolidated Grade School
P.O. Box 160
Salem, WI 53168

**Burns, Stacy** (1)
Richland Elementary School
200 Spell Drive
Richland, MS 39218

**Carroll, Lesa** (5)
Knob Noster Elementary School
405 E. Wimer Street
Knob Noster, MO 65336

**Carter, Terry/Smith, Pam** (6)
Welsh School
2100 Huffman Boulevard
Rockford, IL 61103

**Coleman, Joe F.** (4)
Allensville Elementary School
R.R. 1, Box 668
McArthur, OH 45651

Conley, Sharon (2)
Howard C. Reiche Community
   School
166 Brackett Street
Portland, ME 04102

Dong, Janet Mar (5)
Jean Parker Elementary School
3630 Divisadero Street
San Francisco, CA 94127

Emmons, Marilyn K. (1)
Broadus Elementary
P.O. Box 500
Broadus, MT 59317

Erickson-Danielson, Becky Ann
   (1)
Orchard Lake Elementary School
16531 Klamath Trail
Lakeville, MN 55044

Geren-Saggau, Mary (1)
Arlington Public Schools
Box K
Arlington, NE 68002

Gildroy, Mary Ann (K)
Central Elementary School
600 1st Street W.
Roundup, MT 59072

Goodemote, Shirley H. (1)
Northville Central School
Third Street, Box 608
Northville, NY 12134

Hargett, Charlotte L. (1)
Charleston Elementary
412 Chestnut Street
Charleston, MS 38921

Hartson, Dorothy (6)
Inter-Lakes Elementary School

Laker Lane
Meredith, NH 03253

Herman, Barbara (1, Transition)
Edison Elementary
9th and Main or
134 W. Ninth Street
Bristow, OK 74010-2499

Karna, Faye (K)
Plaza Public
P.O. Box 38
Plaza, ND 58771

Kelly, Sharon (3)
Hopewell Elementary School
35 Princeton Avenue
Hopewell, NJ 08525

Mahloch, Nancy (6)
Edgar Elementary School
Box 248
Edgar, NE 68935

Maloy, Kathryn (2)
Coral Terrace Elementary School
6801 S.W. 24th Street
Miami, FL 33155

Mann, Debra L. (Sp. Ed., K − 5)
Elias Brookings
367 Hancock Street
Springfield, MA 01105

Martin, Helen M. (4)
Fair Street Elementary
695 Fair Street
Gainesville, GA 30505

McCollum, Pauline (2)
Tampa Palms Elementary School
6100 Tampa Palms Boulevard
Tampa, FL 33647

McDonald, M. Faye (K)
West Kemper Elementary School

P.O. Box 250
DeKalb, MS 39328

**More, Leslie** (2)
Republic Primary School
915 E. Highway 21
Republic, WA 99166

**Morgan, Karen** (6)
Whittier School
301 N. 29th Street
Boise, ID 83709

**Parmenter, Barb** (4)
Pine River Area Schools – Tustin
  Elementary
107 Bremer Street
Tustin, MI 49688

**Parnell, Penny** (3)
West Jasper Elementary School
1500 W. 19th Street
Jasper, AL 35501

**Robeson, Betty** (3)
Pangborn Boulevard Elementary
  School
195 Pangborn Boulevard
Hagerstown, MD 21740

**Romano, Wendy** (3)
Walnut Street School
Walnut Street
Woodbury, NJ 08096

**Shepard, Christine** (2)
Red Rock Elementary

685 Millcreek Drive
Moab, UT 84532

**Smith, Kathryn A.** (2)
A. R. Lewis Elementary
1755 Shady Grove Road
Picken, SC 29671

**Smith, Pam/Carter, Terry** (6)
Welsh School
2100 Huffman Boulevard
Rockford, IL 61103

**Spear, Shirley** (L.A., 4, 6)
Wink-Loving Independent School
  District
Box 637
Wink, TX 79789

**Spitzer, Cay** (6)
Patrick Henry Elementary School
1310 Lehmberg Boulevard
Colorado Springs, CO 80915

**Stevens, Linda N.** (1)
Earl Nash Elementary School
Hwy 14, P.O. Box 391
Macon, MS 39341

**Stirler, Marcia** (K)
Janesville Consolidated School
505 Barrick Road
Janesville, IA 50647

**Wood, Daria Ann** (K, Transition)
Jefferson Elementary School
121 N. 5th Street
Riverton, WY 82501

# Secondary Teacher Directory

**Adcock, Ron** (9 – 12)
Theodore Roosevelt High School
4419 Center Street
Des Moines, IA 50312

**Bishop, Elizabeth** (9 – 12)
Mayville High School
6250 Fulton Street
Mayville, MI 48744

**Blunier, Gayle** (10 – 12)
North Posey High School
R.R. 1
Poseyville, IN 47633

**Boyle, Terry** (10, 11)
Robert E. Lee High School
1200 N. Coalter Street
Staunton, VA 24401

**Bryant, Barbara B.** (8 – 12)
New Brockton High School
P.O. Box 429
New Brockton, AL 36351

**Burton, Elizabeth** (9)
Geeter Jr. High
4649 Horn Lake Road
Memphis, TN 38109

**Clark, Ken** (11, 12)
Woodward High School
7001 Reading Road
Cincinnati, OH 45237

**Clemons, Brenda** (10 – 12)
Booker T. Washington High
  School
1111 Park Avenue
Norfolk, VA 23504

**Cullinane, Lynne** (9 – 12)
Hopewell Valley Central High
  School
259 Pennington-Titusville Road
Pennington, NJ 08534

**Daileanes, Charles J.** (8)
Elm Street Jr. High
Elm Street
Nashua, NH 03060

**Eschmann, Bob** (9 – 12)
North Chicago Community High
  School
1717 17th Street
North Chicago, IL 60064

**Fenner, Dan A.** (8 – 12)
Girard High School
1135 Lake Street
Girard, PA 16417

**Gill, Beverly-Ann** (9)
Martin Jr. High School
111 Brown Street
Courage House
East Providence, RI 02914

**Gottsche, Myra M.** (11, 12)
Biloxi High School
1424 Father Ryan Avenue
Biloxi, MS 39530

**Grant, Elaine W.** (9 – 12)
Bangor High School
885 Broadway
Bangor, ME 04401

**Gregory, Frank R.** (9, 12)
Dover High School
625 Walker Road
Dover, DE 19901

**Hungerford, David** (9 – 12)
Central High School
100 Summit Street
Newark, NJ 07102

**Koenigsberger, Thomas**
(Secondary)
Jefferson High School
4145 Samuelson Road
Rockford, IL 61109

**LaPeer, Susan** (9 – 12)
Swampscott High School
Forest Avenue
Swampscott, MA 01907

**Meriwether, Nell W.** (12)
Tara High School
9002 Whitehall Street
Baton Rouge, LA 70806

**Meserve, Howard L.** (11, 12)
Springer High School
8th and County Road
Springer, NM 87747

**Morrow, Kyle Ann** (7 – 12)
Nome-Beltz Jr./Sr. High School

Box 131
Nome, AK 99762

**Naze, Brenda M.** (9 – 12)
Malcolm Shabazz City High
  School
1601 N. Sherman Avenue
Madison, WI 53704

**Neal, Kay** (9 – 12)
Glenns Ferry
545 N. Bannock, Box 850
Glenns Ferry, ID 83623

**Nienkamp, David** (7, 10, 12)
Sandy Creek Jr.-Sr. High School
Rt. 1, Box 127
Fairfield, NE 68938

**Ortman, Jennifer** (9 – 12)
Bridgewater High School
Box 350
Bridgewater, SD 57319

**Pennington, Thomas D.** (6 – 12)
Madawaska Middle/High School
80-7th Avenue
Madawaska, ME 04756

**Puckett, Elaine** (8)
South Paulding Middle School
1200 Nebo Road
Dallas, GA 30132

**Reed, Pamela A.** (9 – 12)
Grand Rapids Central High School
421 Fountain NE
Grand Rapids, MI 49503-3398

**Robinson, Christine** (7)
Winona Middle School
166 W. Broadway
Winona, MN 55987

**Ruby, Nancy D.** (Secondary)
Fair Haven Union High School
Mechanic Street
Fair Haven, VT 05743

**Saberali, Abbas** (7–9)
Memorial Jr. High School
P.O. Box 1409
Eagle Pass, TX 78853

**Schlaak, Kelly** (9–12)
United South Central High School
250 2nd Avenue
Wells, MN 56097

**Sexton, Marilynn** (9–12)
Skyview High School
1775 High Sierra Boulevard
Billings, MT 59105

**Sturgeon, Kimberly** (9–12)
King William High School

Rt. 1, Box 401
King William, PA 23086

**Tanner, Sally** (10–12)
North High School
801 17th Avenue North
Fargo, ND 58102

**Vaughn, Sue** (9–12)
McQueen High School
6055 Lancer Street
Reno, NV 89523

**Worthen, Paul C.** (9–12)
Page High School
P.O. Box 1927
Page, AZ 86040

**Yunker, Robert H.** (10–12)
Northern High School
Route 2, Box 4
Accident, MD 21520

CHARLES M. CHASE, Ed.D. is Director of Alternative Certification for teachers in the College of Education and Social Sciences at West Texas State University. He has a wide and varied background in education, ranging from teaching survival education in the United States Air Force to being assistant dean in the School of Education at the University of Mississippi. He received his doctoral degree in curriculum and instruction from the University of Northern Colorado.

JACQUELINE E. CHASE, M.A. currently owns a business that manufactures teaching aids. She has teaching experience in a variety of educational settings, ranging from preschool to elderhostel. Ms. Chase enjoys volunteering with her sons' many activities. She received her master's degree in recreational administration from the University of Mississippi.

**DATE DUE**

| | | | |
|---|---|---|---|
| | | | |
| | | | |
| | | | |
| | | | |
| | | | |
| | | | |
| | | | |
| | | | |
| | | | |
| | | | |
| | | | |
| | | | |
| | | | |
| | | | |
| | | | |
| | | | |
| | | | |
| | | | |

HIGHSMITH 45-220